Graph Design
for the Eye and Mind

Graph Design for the Eye and Mind

Stephen M. Kosslyn

OXFORD
UNIVERSITY PRESS

2006

OXFORD

UNIVERSITY PRESS

Oxford University Press, Inc., publishes works that further
Oxford University's objective of excellence
in research, scholarship, and education.

Oxford   New York
Auckland   Cape Town   Dar es Salaam   Hong Kong   Karachi
Kuala Lumpur   Madrid   Melbourne   Mexico City   Nairobi
New Delhi   Shanghai   Taipei   Toronto

With offices in
Argentina   Austria   Brazil   Chile   Czech Republic   France   Greece
Guatemala   Hungary   Italy   Japan   Poland   Portugal   Singapore
South Korea   Switzerland   Thailand   Turkey   Ukraine   Vietnam

Copyright © 2006 by Oxford University Press, Inc.

Published by Oxford University Press, Inc.
198 Madison Avenue, New York, New York 10016

www.oup.com

Oxford is a registered trademark of Oxford University Press   •

Library of Congress Cataloging-in-Publication Data
Kosslyn, Stephen Michael, 1948–
Graph design for the eye and mind / by Stephen M. Kosslyn.
p. cm.
Includes bibliograhpical references and index.
ISBN-13: 978-0-19-530662-0; 978-0-19-531184-6 (pbk.)
ISBN-10: 0-19-530662-7; 0-19-531184-1 (pbk.)
1. Graphic methods.   2. Experimental design—Graphic methods.
3. Statistics—Graphic methods.   I. Title.
QA90.K634 2006
001.4'226—dc22   2006010275

9 8 7 6 5 4 3 2 1

Printed in the United States of America
on acid-free paper

Preface

This is a how-to book: how to present information effectively in graphs. Right behind the how is the why: The same principles that govern how our visual systems and minds make sense of the world also govern how we make sense of graphs, and the advice I offer here is rooted in research results. In this book I show that an understanding of the workings of the eye and mind can be used profitably to design good displays.

This book is not just for researchers and designers. It is intended for department managers making monthly reports who are expected to make use of those fancy (or not so fancy) graphics programs in the company computers; for people in any field making presentations; for students. The reader, I assume, is intelligent and interested—but not necessarily possessed of professional training as an academic, a scientist, or a designer. Nor are the applications limited to today's (or tomorrow's) computer programs; your tools may equally well be a drawing board or pencil and paper. This is graph making for the people.

This book is a step-by-step guide to constructing comprehensible graphs. We begin by considering the type of data you have and the general point you want to make, and select the appropriate type of graph. Then we turn to the practical work of producing the display, systematically constructing the overall framework, the content, and the labels, adding color and texture, and reviewing the outcome. The result is a graph that virtually anyone can read immediately, not a puzzle to be pondered.

The reader interested in the details of the principles of perception and cognition that underlie my graph recommendations will find summaries of key findings as well as references for additional reading. Moreover, I have provided three appendices, one that reviews basic concepts in statistics, one that summarizes the psychological principles that motivate my recommendations, and another that provides a checklist of features to consider when evaluating a computer graphics program.

I have many people to thank. The idea for this book was born in 1978 when my then-student Steven Pinker (now a distinguished scientist and scholar at Harvard University) and I responded to a request for grant proposals from the now defunct National Institute of Education. The NIE—through the good offices of Dr. Susan Chipman—wanted to encourage more research on how people comprehend charts and graphs. Steve and I were hired by a company called Consulting Statisticians Inc. (also now defunct) to write a proposal for such research and—to our amazement—won the award. Thus we, along with William Simcox and Leon Parkin, began a three-year research project on this topic. A very large and very technical manuscript was born from this project, which was never published. Steve and I then distilled this into a slightly smaller and no less technical text, also never published, of which several chapters saw the light of day as research treatises.

As time went on, Steve's and my research developed in different directions. Steve investigated the details of how children acquire language and eventually became an internationally known "public intellectual," and I became immersed in cognitive neuroscience and increasingly interested in theories of how the brain functions. Another part of me, however, yearned for something more immediate and concrete. I was particularly interested in a project that would demonstrate that pure research in psychology could pay off in practical ways—and not just in the distant future.

Thus came my first book on graph design, entitled *Elements of Graph Design*, which was published by W. H. Freeman in 1994. That book would never have been written if not for the active support and encouragement of many people. Dr. Susan Chipman stayed in touch with me over the years and kept my interest in the topic alive; I doubt that I would have written this book if not for her. My editor at W. H. Freeman and Company, Jonathan Cobb, was absolutely superb; I cannot thank Jonathan enough for his patience, his thoughtful suggestions and corrections, and his constant encouragement. Megan Higgins, art director at Freeman, was so enthusiastic about this project that I was driven to refine my presentation far beyond my original expectations. After I thought I was finished, Nancy Brooks began to edit the book in earnest, and developed such good ideas that I ended up reorganizing it almost from scratch. Nancy read and thought about every word and stared at every graph; if there was any chance of an ambiguity, Nancy showed me how to eliminate it. Nancy is one of those people with rare vision, able to see the diamond in the rough—and able to see how to cut it and polish it. Like most diamond cutters, she needed just the right touch and lots of patience, and I thank her for both.

Elements of Graph Design received the best reviews of any book I've ever written, but—because of a series of management changes at the publisher—received no advertising (none, zero). This did not do wonders for sales. In 2004 *Elements* went out of print, and rights reverted to me. At that point I decided to use *Elements* as the foundation for the present book, which differs from the previous one in five major ways: First, I completely revised and recast the principles that underlie the recommendations. The principles I used in *Elements* were not as well organized as they could have been, and hence were unnecessarily difficult to use and remember. I have reorganized the psychological material in the previous book, preserving what was good but making the material easier to grasp and retain. Second, I have read all the studies I could locate on graphic communication since 1994, and used the new material to sharpen the recommendations (when appropriate) and also to update the reviews in the text. Third, I have integrated the reviews of the relevant science of perception and cognition, as well as the results of studies of graphics per se, into the text, rather than tuck this information into endnotes (which were often overlooked in *Elements*). Fourth, I have corrected many errors and clarified throughout. I am particularly grateful to Ronnie Lipton for her help on this score; her detailed comments improved this book immensely (and the book would no doubt be noticeably better if I had had the time to follow all of her recommendations and suggestions—she's a real pro!). Finally, the book has been thoroughly redesigned, with the material flowing down the page rather than requiring visual zigging and zagging. I thank medical illustrator extraordinaire Steven Moskowitz for his constructive critique of *Elements*, which has improved the design of this book.

This book would never have come into the light of day if not for Catharine Carlin at Oxford University Press. I wish to thank her for her faith in this book and help at every phase of the project, and also thank her assistant Danny Bellet for his patience. In my efforts to "get it right," I no doubt surprised (and dismayed) Catharine and Danny with the waves of revisions I kept turning in. But they bore this with good humor and never failed to respond with useful advice. I also thank Lisa Stallings for managing what turned out to be a surprisingly complex production process, and Will Boehm and Tracy Baldwin for patience and good sense in designing the layout of the book.

I am also grateful to Patrick Cavanagh, Lynn Cooper, Reid Hastie, Richard Meir, and Armand Schwab for unusually helpful comments on an earlier version of the manuscript, and to Charles Stromeyer for pointers into the technical details of research on visual perception. Christopher Chabris, Gregg DiGirolamo, Anne Riffle, Lisa Shin, Andrew Stultman, and Amy Sussman provided all manner of technical assistance. I am particularly grateful to Chris and Andrew for tracking down so many references and interesting examples of good and bad graphics, and for giving me solid advice and suggestions about how to present this material. And I am grateful to Anne and Lisa for their unflaggingly cheerful management of the myriad details of this project. Chris, Lisa, Bill Thompson, and Adam Anderson helped find the data that are presented in the graphs I use as illustrations, and I thank them for their efforts.

And, of course, I must thank the funding agencies that made this project possible. First, the John Simon Guggenheim Memorial Foundation awarded me a fellowship, which gave me time to work. Second, the conceptual foundations and new empirical findings described herein were supported by the National Science Foundation (NSF) and the National Institutes of Health (NIH). Specifically, I am obliged to note that this material is based upon work supported by the NSF under grant 0411725 and NIH under grant 2 R01 MH060734-05A1. Any opinions, findings, and conclusions or recommendations expressed in this material are those of the author and do not necessarily reflect the views of the NSF or the NIH.

Finally, I thank my son David for repairing and redrafting dozens of illustrations from *Elements*; his skills with Photoshop must be seen to be believed. And I thank my son Justin for his critical comments and advice about the principles I reformulated; his facility with formalisms is remarkable. And I wish to thank my wife, Robin Rosenberg, and children, Justin, David, and Nathaniel, for their patience and good humor as I finished this book immediately on the heels of two previous ones.

Contents

How to Use This Book

To get the "big picture" of the many ways in which principles of perception and cognition lead to specific recommendations about graphs (and other displays), read the book straight through, beginning to end. To use it as a workbook or manual to guide you in the design and construction of a particular display, follow the signposts provided in the text.

In either case, begin with chapter 1, where we see how principles of perception and memory are a two-edged sword: If they are ignored, a display can be uninterpretable; if respected, it can be read at a glance. Move on to chapter 2, which will guide you to the best choice of display format for a particular purpose. If you are using *Graph Design for the Eye and Mind* as a workbook, be guided by the paragraph at the end of each chapter called "The Next Step." Depending on the format you have chosen, "The Next Step" in chapter 2 will direct you either to chapter 3 (to make an L- or T-shaped framework) or to chapter 4 (if your choice is, say, a pie graph or visual table). Now you are beginning the work of producing the graph. If your graph has an L- or T-shaped framework, "The Next Step" at the end of chapter 3 will then send you either to chapter 5 or to chapter 6 to continue the process. Most recommendations are illustrated with **don't** and **do** examples; if you are willing to trust me, you can simply read the recommendations and look at the pictures and be guided accordingly.

Having constructed your graph in outline, you are ready to consider further refinements. Chapter 7 provides recommendations for using color, hatching, shading, and three-dimensional effects, as well as for constructing inner grid lines, background elements, keys, and captions. This chapter also provides recommendations for creating multipanel displays.

Chapter 8 outlines how people can lie with graphs—warnings for us all. After the display has been created, the recommendations in chapter 8 can be used to check its overall appearance to ensure that it accurately conveys the patterns in the data. Finally, chapter 9 discusses how the principles can be used to create effective charts, maps, and diagrams and offers a brief look at the possibilities for better information displays that are being provided by new technologies.

An important note: The recommendations offered in this book are guidelines, not hard-and-fast rules. In some cases, aesthetic concerns, context, or your intended emphasis may override my advice. Moreover, the recommendations occasionally may interact in unexpected ways. It is sometimes helpful to generate a rough sketch of several alternative displays that respect the recommendations and evaluate each one. This is particularly easy to do on a personal computer with a good graphics program. Cleveland (1985) stresses the idea that graph production is an iterative process and suggests regraphing data repeatedly and deciding on the final design after viewing the alternatives. This is reasonable advice if you have large amounts of data that can be organized in various ways, or if you want to explore the utility of alternative display formats that are appropriate for that type of data. However, as you master the psychological principles (which provide the foundations for all my advice), you will find that you can rely increasingly on your intuitions about good display design, without needing to try out many alternative versions of your display.

Graph Design
for the Eye and Mind

Looking With the Eye and Mind

A PICTURE CAN BE worth a thousand words—but only if a reader can decipher it. Pictures may be hard to fathom when they are too small or blurry; they are also meaningless, or nearly so, when their content is organized in a way that we cannot easily comprehend. The worst offenders may be graphs, which are pictures intended to convey information about numbers and relationships among numbers.

Graphs have become a pervasive part of our environment; they appear in magazines and newspapers, on television and on cereal boxes. Given their near ubiquity, it is surprising that so few graphs communicate effectively. One reason for the abundance of bad graphs is the proliferation of low-cost personal computers and "business graphics" programs, which often seduce the user into producing flashy but muddled displays. (When was the last time you saw an advertisement for a personal computer whose screen did not feature an impressive display?) But although the ease of creating charts and graphs is a major selling point for personal computers, one rarely hears anything about the utility of the displays the machines produce. (When was the last time you could figure out what the display was supposed to mean?) Johnson, Rice, and Roemmich (1980) reviewed the graphs in annual reports of fifty Fortune 500 companies and found that almost half the reports contained at least one graph that was inappropriately constructed. (However, the fact that the most common errors distorted recent trends suggests that not all errors could be attributed to simple ignorance of how to make good displays!) Computers, of course, do not create the problem; they merely multiply it. Confusion and lack of clarity are apparent also in many hand-drawn and traditionally produced graphs.

The obvious question is, Why? Why this muddiness in so many instances of such a common form of presentation of information? The answer is the thesis of this book: Many graphs are designed without consideration of principles of human visual perception and cognition. Even most how-to guides for creating charts and graphs ignore the obvious fact that the intended audience consists of human beings who by their nature have specific perceptual and cognitive strengths and limitations. There is a wealth of important research about the way we perceive and reason, but it has seldom been mined to aid in the design of good visual displays. One notable exception is the excellent book written by Cleveland (1985). However, Cleveland's orientation differs from mine in the following ways: He does not adopt a step-by-step approach, offers fewer specific recommendations, discusses a wide range of exotic and special-purpose displays, often offers mathematical treatments of the material, and often focuses on uses of graphs in data analysis. Cleveland is a statistician and has a statistician's sensitivity to the nuances of data; I highly recommend this book, and the book by Chambers, Cleveland, Kleiner, and Tukey (1983), to anyone interested in using graphics to analyze data.

A graph is a visual display that illustrates one or more relationships among numbers. It is a shorthand means of presenting information that would take many more words and numbers to describe. A graph is successful if the pattern, trend, or comparison it presents can be immediately apprehended. Our visual systems allow us to read proportional relations off simple graphs as easily as we see differences in the heights of people, colors of apples, or tilts of pictures mishung on the wall. We are visual creatures and are good at noting relative differences in sizes, lengths, orientations, and some other visually perceptible properties. Graphs can also allow us to appreciate the quantitative relations among multiple elements and thus can provide us with precise information.

The goal of this book is to help your graphs do their job, which is to communicate effectively. Graphs are used for two major sorts of purposes. On the one hand, they are used to help you understand what your data are telling you. Consider a remark made by the statistician Tukey (1977, p. vi), to the effect that a good pictorial display of data "*forces* us to notice **what we never expected to see**" (italics and bold in the original). On the other hand, they are used to convey to others what the data mean. For present purposes I would say that a good graph forces the reader to see the information the designer wanted to convey. This is the difference between graphics for data analysis and graphics for communication. Given the present goal, I avoid novel or exotic types of displays and focus on the types that are used most commonly in business and communication.

Whereas a graph used solely to analyze data need not be attractive, a graph used to communicate should be visually inviting. Making a display attractive is the task of the designer, whose talent and visual sense give the graph snap and visual appeal. (For many examples of visually interesting graphs, as well as some guidelines for designing them, see Bertin, 1983; Holmes, 1984; Tufte, 1983, 1990; for a review of five books on graph design, see Kosslyn, 1985.) But these properties should not obscure the message of the graph, and that's where this book comes in. The recommendations offered here will enable you, step by step, to construct the essential elements of an effective display, but the designer's creativity will have plenty of leeway. The recommendations are grounded in facts about the workings of our eyes and minds that affect the way we take in and process visual information (note: I use the term "mind" to refer to "what the brain does," to brain function—so, when I refer to the mind, I'm also implicating the brain). The principles I develop are firmly rooted in these facts, and these principles give rise to my recommendations. A graph designed according to these recommendations plays to the perceptual and cognitive strengths of those who see it and avoids being defeated— that is, misunderstood—by the inherent weaknesses of our perceptual and cognitive systems.

Psychological Principles of Effective Graphics: The Eight-Fold Way

The recommendations I offer here are rooted both in findings about how people perceive and comprehend visual displays per se and in many facts about how our eyes and minds generally organize and interpret the world around us; indeed, most of the findings I draw upon are now considered classics (for overviews of the psychology of perception, see Bloomer, 1990; Dodwell, 1975; Frisby, 1980; Gregory, 1966, 1970; Hochberg, 1964; Kaufman, 1974; Osherson, Kosslyn, & Hollerbach, 1990; for overviews of cognitive processes, see Anderson, 2004; Glass & Holyoak, 1986; Kosslyn & Koenig, 1995; Osherson & Smith, 1990; Smith & Kosslyn, 2006). The advice offered in this book is derived from a set of overarching psychological insights, which can be summarized by eight principles. Each of these eight principles will come into play, singly or in combination, as we create different kinds of graphs in the following chapters. I think of this approach as the Eight-Fold Way to articulate graphics. (These principles are summarized in appendix 3.)

The eight principles are organized into three sets because they play special roles in accomplishing different goals. Specifically, any good graphic should allow you to (1) connect with your audience, (2) direct the reader's attention through the display, and (3) promote understanding and memory. In what follows, I introduce the principles in the context of these goals, with the understanding that all of the previously discussed principles bear on achieving each subsequent goal, in addition to the principles introduced in the section.

Connect With Your Audience

Information is communicated effectively only when it focuses the readers' attention and interest on a specific message. Graphs and other sorts of visual displays serve to help readers answer a particular question, be it very specific (e.g., whether girls typically are taller than boys in early adolescence; whether the dollar buys more in China than in the U.K., etc.) or relatively general (e.g., how much office space per square foot costs in different cities; what was the exchange rate between the U.S. dollar and Euro over the past five years). Two principles will help you to reach your readers.

The Principle of Relevance
Communication is most effective when neither too much nor too little information is presented.

The first thing you need to do when beginning to prepare a graphic is to decide on exactly what message you want to convey. This is crucial because readers expect to see all and only the relevant information. Presenting too little information will puzzle the readers, and presenting too much will overwhelm them with needless detail.

The Principle of Appropriate Knowledge
Communication requires prior knowledge of relevant concepts, jargon, and symbols.

You also need to take into account the nature of your audience. "Know your audience" is rock-solid advice. To communicate effectively, a display must be pitched at the right level for the readers you wish to reach. This is true in terms of both the type of display you use (e.g., a standard bar graph is familiar to everyone, but a box-and-whisker chart is not universally understood) and the specific information included (e.g., everyone understands amount, but not everyone knows what a first or second derivative is). Be sure that your intended readers will understand the concepts, symbols, and jargon you use. In addition, be sure that the readers will have the appropriate background knowledge to make sense of your message. A good display presents new information—but the information should not be so new as to be completely unrelated to anything else the readers know. If you assume that the readers know too much, you will rely on concepts, facts, or conventions that are unfamiliar to them—and you may as well be speaking in tongues. If you assume that they know too little, you'll overstate what they find obvious.

Direct and Hold Attention

A good display draws in the readers and eases them through the display, helping them to notice first what's most important and then to make all the appropriate distinctions. Three principles will help you accomplish this goal.

The Principle of Salience
Attention is drawn to large perceptible differences.

The most visually striking aspects of a display will **draw attention** to them (as did this bold type), and hence they should signal the most important information. All visual properties are relative, and thus what counts as "visually striking" depends on the properties of the display as a whole.

The Principle of Discriminability
Two properties must differ by a large enough proportion or they will not be distinguished.

Perhaps most fundamentally, to ensure that the reader pays attention properly, two visual properties must differ by a large enough proportion or the reader will not be able to tell that they are different (and hence cannot pay attention to one vs. the other). Notice how much harder it is to tell the difference between the members of the first pair below compared to the second:

> *m* versus *rn*
> *m* versus *o*

Detectability is special case of this aspect of the principle: Marks must be large or heavy enough to be noticed (i.e., distinguished from the background). You probably will have problems with the last part of this sentence.

▬ The Principle of Perceptual Organization
People automatically group elements into units, which they then attend to and remember.

It is natural to assume that our eyes are simple receiving systems that, like a TV camera, register the world as it is. A moment's reflection should convince you that this analogy is misleading. Consider the following observations: We readily notice a gain or loss of five pounds on a thin person but may have to strain to see it on someone obese; the box of one brand of soap may look almost twice as big as a competitor's but contains nowhere near that much more soap; a row of reflectors on a dim highway or a formation of geese flying overhead is seen as a single pattern, not as individual objects; former gymnasts watching Olympic gymnastics see much more than do people who are watching the sport for the first time. These phenomena are consequences of the way our eyes and mind work. The mind is not a camera. We are not simply passive receptors; we actively organize and make sense of the world, and when we do so we are at the mercy of the wiring of our eyes and brains.

The Principle of Perceptual Organization has the following distinct aspects.

Input Channels. Although many aspects of the ways we unconsciously organize what we see are obvious once they are pointed out, some are not. For example, consider the illustration on the opposite page. If you wear glasses, try taking them off and looking at it; if you do not wear glasses, try looking at it from about ten feet away. You will find it easier to identify the subject as Abraham Lincoln when the picture is out of focus. Why?

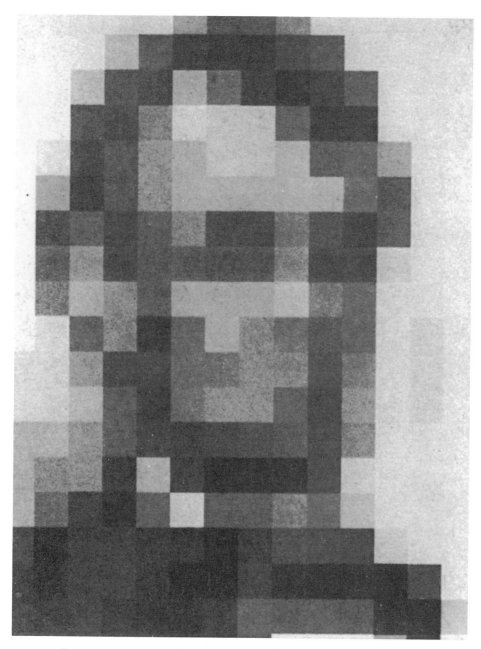

Figure 1.1. A picture of a president of the United States. To see it more clearly, take off your glasses or step back about ten feet from the page.

When you look around, it seems that you see everything at roughly one level of sharpness, just as would a camera with one lens. But in fact our visual system operates as if it has a number of distinct "input channels." These operate like a set of different lenses, each of which is adjusted to register a different level of detail. Some channels are sensitive only to relatively large changes in a pattern, whereas others are tuned to fine details.

The Lincoln figure was created from an image like that on a television screen, which is composed of tiny dots of varying degrees of lightness or darkness. The lightness values of the dots were then averaged within each square, making each square a uniform shade. Although the averaging process masks fine variations in the pattern (e.g., the hairs of Lincoln's beard, his nostrils, etc.), the edges of the imposed squares introduce sharp changes where none previously appeared. Defocusing the picture will not remove any of the actual details of the picture; these were already removed by the averaging process. But defocusing will remove the details caused by the edges of the squares. The useful information from the coarse visual input channels (the overall outlines) was being obstructed by inappropriate information from the fine-detail channels (the edges of the squares); blurring the picture prevents these detail channels from providing spurious input—thus unmasking the information transmitted by the coarser channels. As a result, you see the image better when the picture is blurred (for more information on this phenomenon, see De Valois & De Valois, 1988; Julez, 1980).

Why does our visual system operate at numerous levels of acuity at once? Why doesn't it just use the single "best" level? The answer is that there isn't one. Which channels are optimal depends on the task at hand. For counting hairs, the best level of acuity would be the highest; for counting cows, the best would be a lower one, which allows us to ignore extraneous detail.

What do input channels have to do with visual display design? The crucial fact is that a viewer cannot help but pay attention to adjacent patterns if they are processed by the same channel. If you want readers to pay attention at first glance to specific bars, wedges of a pie, areas on a map, and so on, the patterns should be so constructed that they are processed by different channels; if patterns are processed by the same channel, readers will have to work hard to pay attention to individual items. Researchers have measured the variations in light and dark that lead patterns to be registered by different channels, and their conclusions—discussed in chapter 3—can be helpful guides to making effective use of cross-hatching, patterns of dashes in lines, and so on.

Three-Dimensional Interpretation. Our eyes and minds attempt to interpret visual patterns as signifying three-dimensional objects whenever possible. Specifically, the mind operates as if it unconsciously assumes that shapes have regular, closed contours composed of parallel and perpendicular sides, that objects have the same shading and markings all along their surfaces, and that the world is made of cohesive surfaces rather than clouds composed of tiny drops or shards. Thus, we often will organize lines on a page (or screen) as if they signified the presence of three-dimensional objects. However, two-dimensional patterns of lines are not actually three-dimensional objects, and (as painters through the ages have discovered, to their regret) such patterns do not perfectly convey the third dimension (Stevens, 1974, 1975).

Here is another example of this aspect of the Principle of Perceptual Organization. The brain senses depth in part by exploiting the slightly different images that are registered by each eye. By analogy to a similar process in hearing, this is called stereo vision. In hearing, the brain reconciles the slight differences in the time a sound arrives at each ear in order to localize the source of the sound; in vision, it reconciles slight left/right disparities in the images in each eye to infer the distance of the object. Stereo vision works well, but it can be tricked to produce an interesting illusion: We see warmer colors (e.g., red) as "closer" to us than cooler colors (e.g., blue; for more information on this phenomenon, see De Valois & De Valois, 1988; Julez, 1980). The illusion occurs because light waves have different wavelengths, which we see as different colors. You can separate out the different colors in ordinary light by holding a prism in front of a window and observing the rainbow that is projected on the wall. The lens of the eye acts like a prism because the eye is aimed slightly to one side when we look straight ahead, and so the lens is angled. Light of different wavelengths is projected to slightly different locations on the retina of the eye just as it is projected to slightly different locations on the wall when we hold up a prism. The result is a "false stereo" effect, where long-wavelength, warmer colors seem "closer" to us than short-wavelength, cooler colors. This information leads me to recommend that when two lines cross (as in a line graph), the warmer one should overlap the cooler one. If it does not, the back line will seem to be fighting to come forward, trying to snake around the one in front —which is distracting. Moreover, for the same reason cool colors should be used for the background, and warm colors for the foreground (e.g., for text, graphs, etc.).

Integrated Versus Separated Dimensions. In addition, some visual dimensions are automatically integrated together. For example, when you pay attention to the height of a rectangle you will also pay attention to its width. Great effort is required to register changes on one integrated dimension while ignoring changes on another. Such *integral dimensions* cannot be used effectively to convey different types of information. In contrast, *separable dimensions*, such as the length of a radius of a circle and its angle, can be used effectively to convey different measurements.

Grouping Laws. Another way in which the mind organizes visual elements is by grouping them. There are a number of so-called "laws of perceptual organization" that describe how we group the objects before us. For example, marks near each other will tend to be grouped together (which is known as the *law of proximity*). For example, xxx xxx is seen as two groups, whereas in xx xx xx the same number of elements is seen as three. Thus, depending on how bars in a bar graph are spaced, they are more or less likely to be seen as grouped. Bars in the same group are readily compared. As we shall see, these laws are especially important for associating labels with scales, bars, wedges, lines, and so on.

Similarly, marks that suggest a line, even a dashed or dotted one, will tend to be grouped together (*law of good continuation*). For example, — — — — — — is seen as a single unit, not six separate ones; on the other hand, — — — __ __ __ is seen as two units. Bars that are arranged in order of increasing or decreasing size (so that their tops can be grouped via good continuation) will be more easily apprehended than bars arranged in a way that violates good continuation. And similar marks will tend to be grouped together (*law of similarity*). For example, OOOXXX is seen as two groups. These grouping laws can often be used to help readers to pair corresponding bars, lines, or regions when more than one comparison is being made in a single graph.

Two of these grouping laws will be particularly important in the recommendations I provide: The *law of common fate* specifies that lines or marks that seem to be headed in the same direction will be grouped together. Compare the two line graphs below. A display containing parallel lines is far easier to understand than one containing nonparallel lines; the parallel lines will be grouped into a single unit, whereas the same number of non-

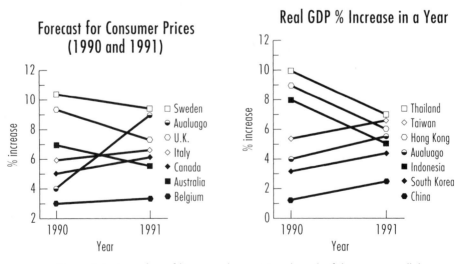

Figure 1.2. A number of lines can be registered easily if they are parallel, forming a single perceptual group. The difficulty of reading the display increases as more perceptual groups—not more lines—are present.

parallel lines would be seen as individual units. This grouping law will play a critical role when we consider how to present complex sets of data in multiple panels in chapter 7. Finally, regular enclosed shapes are seen as single units: [ ] is seen as one unit, but [— is not. Line graphs sometimes produce patterns that are organized perceptually in accordance with this *law of good form.*

These grouping effects can be very powerful. Take a look at a familiar pattern, the Star of David. Good continuation, proximity, and good form lead the mind to organize the lines into the two overlapping triangles that we readily see (Reed & Johnsen, 1975). But there are other geometric figures embedded in the design of the star which are not obvious without expending effort to inspect it carefully. For example, to find the three parallelograms in the star, we must consciously look for them, tracing individual lines to find their perimeters. The parallelograms are no less present in the design than are the triangles, but they are much more difficult to see because no grouping effects, in which eye and mind join forces, are in play to help us see them at a glance.

These grouping effects sometimes require adding lines (e.g., to create symmetry). To obtain them, you will sometimes need to violate Tufte's (1983) recommendation that a designer eliminate all ink that does not convey information—even if this results in the elimination of symmetry, closure, and other simplifying properties. Tufte calls extra ink "chartjunk" and computes a "data:ink" ratio: the amount of ink used to convey data compared to total ink used in the display. Graphs that have a low data:ink ratio he terms *boutique graphics.* However, Carswell (1992a), in her review of the literature on graph reading, found little support for the importance of a high data:ink ratio. As Spence (1990) points out, contrary to this advice, more ink may allow people to read displays more quickly in some circumstances. For example, in one of his experiments Spence found that participants could compare two boxes faster than two vertical lines. As a general rule, additional ink should be helpful if it completes a form, resulting in fewer perceptual units.

Figure 1.3. The three parallelograms embedded in the Star of David.

Promote Understanding and Memory

To communicate effectively, your display should be understood at a glance and later recalled without effort. If you exploit key facts about perception and cognition, you can achieve this goal. If you instead ignore such facts, you can inadvertently tax the readers' information processing abilities to the point where the display is confusing and overwhelming. The following principles—in conjunction with those already discussed—will help you to promote good understanding and subsequent memory of your display.

The Principle of Compatibility
A message is easiest to understand if its form is compatible with its meaning.

The injunction *not* to judge a book by its cover is an attempt to fight our natural tendency to do just that; we take appearance as a clue to the reality. For example, for many Americans, an upper-class English accent seems to add about ten IQ points to a speaker and makes what he or she says more persuasive. Just as we can be persuaded by intellectual or verbal appearance, so we are influenced by visual appearance.

Many researchers have stressed that the structure of an effective display must lend itself to being used in the appropriate ways. In fact, there is a long and venerable tradition of studying ways in which different types of displays are useful for different purposes (e.g., DeSanctis, 1984; Friel et al., 2000; Washburne, 1927). For example, Sparrow (1989) argued that graph designers must consider the *task/display compatibility* and emphasized that a good display should directly contain the information needed to perform the task at hand and should not require the viewer to transform the display mentally or to compute new relations. Along the same lines, Tversky, Morrison, and Betrancourt (2002) proposed a "congruence principle," which states that "the structure and content of the external representation should correspond to the desired structure and content of the internal representation" (p. 3). Vessey (1991) and Umanath and Vessey (1994) make a similar point in discussing the concept of "cognitive fit" between the type of display and type information to be conveyed. In addition, Wickens and Carswell (1995) developed a particular version of this idea, which they call the "proximity compatibility principle"; this principle stresses that integrated displays, such as line graphs (in which distinct data points are not shown, but rather are embedded in a continuous line), are best suited for tasks that require integrating information, whereas more separable displays, such as bar graphs, are best suited for tasks that require accessing discrete data points.

The Principle of Compatibility, as I formulate it, has several related aspects, as follows.

Surface—Content Correspondence. In a very real sense, what the reader sees is what the reader gets. A famous demonstration of what happens when this principle is violated was given by John Ridley Stroop in 1935. Stroop showed people the names of colors written in different colors of ink: For example, the word "red" was written in red, blue, or green ink; the word "blue" in red, blue, or green ink, and so on. When participants were asked to report the color of the ink, they took more time and made more errors when the word named another color than when it named the color of the ink. Similar interference occurs when people are asked to read the words "large" and "small" written in small and large typeface, respectively. We are impaired when the two messages, that from the physical stimulus itself and that from the meaning, conflict (for reviews, see MacLeod, 1991; Smith & Kosslyn, 2006). The brain attempts to fit all inputs into a single coherent pattern and balks when there is conflict. A graph is no place for the Stroop phenomenon.

Similarly, a continuous rise and fall of a line will naturally be taken to reflect a continuous variation in the entity being measured. If the changes in that entity are in fact not continuous but discrete, the continuity implied by a line graph can be misleading; a bar graph would better represent the actual situation being depicted.

More Is More. Perhaps the most fundamental implication of the Principle of Compatibility is the observation that "more" of something in a display should correspond to "more" of a substance—higher bars or lines, larger wedges or regions, or the greater extent of a surface should indicate more of the measured material. I once saw an egregious example of a violation of this principle: The artist wanted to convey the burglary rates of different cities and had constructed a variant of a horizontal bar graph; each bar was a row of small pictures of houses, each house representing a fixed number of houses in that city. In each bar only one house was shown being hit by a burglar—which meant that the *longer* the bar, the *lower* the rate. At first glance, one saw the longer bars as meaning "more" and had to fight the inclination to see "more as more" to realize that the longer bars actually indicated "less."

Perceptual Distortion. Some visual dimensions are systematically distorted; notably, area, intensity, and volume are progressively underestimated as they increase. In contrast, although line length is registered relatively accurately, vertical lines appear longer than horizontal ones of the same length. For instance, take a look at the top hat on the following page; it is as

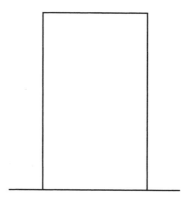

Figure 1.4. The top-hat illusion. Measure the height and the brim of the hat to convince yourself that their extents are the same.

wide as it is high, even though it appears higher. Our visual systems are prey to various illusions—we see properties that are not present (e.g., bent lines when straight ones exist; see Frisby, 1980; Gregory, 1966; Robinson, 1972). In fact, some illusions are specific to graphs. For example, Tversky and Schiano (1989) and Schiano and Tversky (1992) showed participants a pattern that resembled an abstract line graph, namely, an L-shaped bracket that contained a line. They asked the participants later to reproduce the slope of the line from memory, and found that when the participants were told that the stimulus was a graph, they later misremembered the line as being closer to forty-five degrees than it actually was. This effect depended on the participants' believing that the stimulus was a graph, not simply a visual pattern. (For discussion of a number of different illusions in graphs, see Graham, 1937; Kolata, 1984; Poulton, 1985; Schiano & Tversky, 1992; Tversky & Schiano, 1989).

Spatial Imprecision. In addition, some distortion arises because of the inherent *imprecision* of the visual system. Notably, shapes and locations are not precisely conjoined during perception. Imprecision in judging spatial relations apparently occurs because different brain systems are used to register object properties (e.g., shape, color, and texture) and spatial properties (e.g., location, size, and orientation).

Cultural Conventions. The appearance of a pattern should be compatible with what it symbolizes. This is true even when the meaning arises from common *cultural conventions*; for example, "red" means "stop" and green means "go," in Western cultures, and thus printing the word "stop" in green and "go" in red would violate this principle.

The Principle of Informative Changes
People expect changes in properties to carry information.

The reader will interpret any change in the appearance of a display (changing the color or texture, adding, or deleting lines, etc.) as conveying information. By the same token, the reader will expect any piece of information that should be conveyed in a display to be indicated by a visible change in the display. For example, a mark should clearly indicate where a scale has been truncated; a change from current to projected data should be clearly indicated; all important components of a display typically should be labeled.

Principle of Capacity Limitations
People have a limited capacity to retain and to process information and will not understand a message if too much information must be retained or processed.

The brain is part of the body, and like any other organ it has its limitations. The Principle of Capacity Limitations has two major aspects, as follows.

Short-Term Memory Limits. We can keep only a certain amount of information in mind at any one time. A graph (or any information display, for that matter) should not require the reader to hold more than *four* perceptual groups in mind at once. G. A. Miller (1956) originally suggested that we could hold seven groups (which he called "chunks") in mind, but later work has shown that the number is more like four (e.g., Ericsson, Chase, & Faloon, 1980). Observe the pairs of displays to the left and notice how much harder it is to tell whether the ones on the left are identical to the ones on the right as you progress down the column of pairs. Also notice how you start to try to organize the lines into groups, using the grouping laws just described, after the fourth or fifth row in the figure.

Figure 1.5. In each pair, is the pattern the same or different? Notice how the decision becomes more difficult as you move down the page.

One of the reasons graphs are useful is that they help to circumvent limitations of human information processing. Although we can consider only about four groups at a time, we can absorb much more information if it is translated into visual patterns. The table below provides data on blood levels of the (imaginary) fat parafabuloid in men and women at two age and income ranges. Is there one group (defined by a combination of age, gender, and income) that shows an unusual trend—a tendency for something to increase or decrease—over age? Now look at the graph titled "Parafabuloid' Level for Age, by Sex and Income." Here it is apparent that all groups but one show lower levels of "parafabuloid" with increased age—women in the lower income group have the reverse trend. Spotting this trend in the table is difficult; in this graph, it's easy: Differences in the orientations of the lines convey the different trends, and the eye and mind quickly register such differences.

"Parafabuloid" level by Income, Sex, and Age

Income Group	Males		Females	
	Under 65	65 or over	Under 65	65 or over
$0–24,999	250	200	375	550
$25,000+	430	300	700	500

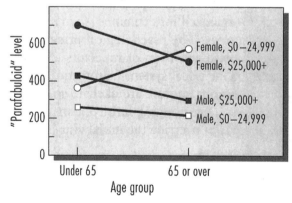

Figure 1.6. The contrasting slope of one line makes the odd group easy to spot; no such visual cue can be given in a table.

Processing Limits. But not any graph will do. In the graph headed "Parafabuloid' Level for Sex, by Age and Income" the same data are displayed but arranged differently. In this graph the important conclusion—that there is one group that shows a trend different from the others—is buried. The distinctions are not shown by the slopes of the lines but by differences in the relative differences of the heights of the symbols. Our visual systems easily register differences in slopes but do not register differences in differences of heights well—and hence the information in this second display is not easily taken in and compared. To be effective, the processing operations required to decode the display must not exceed our human capacities. As you lay out a graph, you must decide which is the most important part of the data; it is that variable that should appear on the horizontal axis, because then the differences in the effects of this variable will correspond to differences in the slopes of the lines or variations in the progression of bars produced.

In addition, our ability to read graphs is affected by other sorts of processing limits. For example, Meyer, Shinar, and Leiser (1997) observe that one consequence of increased complexity is that the viewer must engage in more visual searching, which requires effort. We see only a small region with high acuity and therefore must move our eyes to scan a larger region. And in the absence of knowledge to guide us directly to a key spot, we must explore what is put before us—and thus the more there is, the

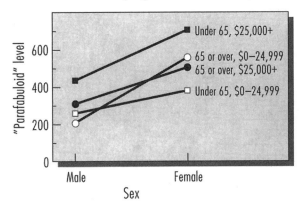

Figure 1.7. This arrangement of the data makes it hard to detect the different trend shown by one group.

more difficult such exploration will be. It is a psychological, not a moral, fact that people do not like to expend effort and often will not bother to do so, particularly if they are not sure in advance that the effort will be rewarded. If you expect readers to extrapolate a trend from a bar graph, you are making them do extra work by having them connect the tops of the bars in their mind's eye to produce a line they can follow. Why not give them a line graph in the first place? On the other hand, if you want to point out specific values, don't use a line that your readers will have to break up mentally into points; give them a bar graph, which indicates specific values directly. In both these cases, the selection of a graph type unsuitable for the data means more work for the reader and an increased likelihood that your graph will fail to convey your message.

The Principles in Action: An Illustration

Look at "Nutritional Information per Serving," a horrendous example of overtaxing the reader's perceptual and cognitive systems (and patience). At one point federal regulators considered requiring such displays on food packaging. Try to read this display. Some people never are able to understand it, even if they keep at it; most give up after a first glance.

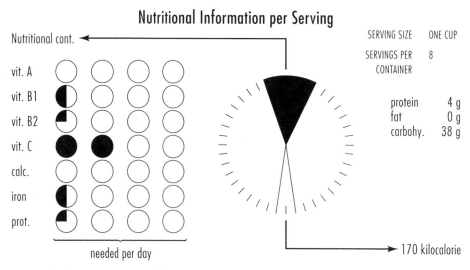

Figure 1.8. A complex display intended to indicate the nutritive content of food. Try to decipher it.

The problems are many. First, the reader is likely to interpret the small circles in the left panel as pie graphs (their form taps into our knowledge of such displays). Pie graphs divide a single whole into its proportions, but each of these circles does not correspond to a single whole. Rather, the total of four circles in each row is supposed to depict a whole. Unless the readers are familiar with such displays, this device violates the Principle of Appropriate Knowledge.

This arrangement is particularly unfortunate because the circles are positioned closer to the ones above them than to the ones to the sides, and so the grouping principle of proximity leads the viewer to see them as organized into columns; in fact, they should be seen as organized into rows. The Principle of Perceptual Organization has been ignored, with the result that the readers must try to override what they see automatically. Moreover, the same length row is used in all cases, which violates the Principle of Compatibility; we tend to see more as more and notice the length of rows before we notice the amount of material in the assembled circles.

Now look at the center panel, which is supposed to convey two different sorts of information: the overall proportion of "Nutritional cont." (actually, vitamins and minerals; the black wedge) and the number of calories (the white wedge) per serving. The Principle of Informative Changes has been violated here: The label is insufficient. Moreover, the Principle of Appropriate Knowledge is again violated because this part of the display uses the technically correct—but less familiar—term "kilocalorie" (which also should have been the plural "kilocalories") to mean "calorie."

It is also unfortunate that the wedges line up to form a single shape; the Principle of Perceptual Organization will lead the readers to think that the two wedges are related, but they are not. In addition, there's a problem in that the panel looks like the familiar pie graph, possibly misleading the reader into thinking that the two wedges are different proportions of the same whole. The absence of labels on the arrows forces the reader to work to figure out that they correspond to different things: "composed of" and "produces" for the left and right arrows, respectively; this is a violation not only of the Principle of Informative Changes but also of the Principle of Capacity Limitations—any time the viewer must work to figure out the meaning, the communication is in danger of failing.

But an even more severe violation of the Principle of Capacity Limitations is evident: To determine the actual amounts, the reader must count 40 tick marks around the circle (mentally supplying the missing ones); a tedious and time-consuming exercise. The fact that specific multiples of tick marks are important would have been more discriminable if every 10th one were bold. Thus, even the Principle of Discriminability has been violated here.

The confusion continues. Notice that "protein" ("prot.") appears in the leftmost portion of the display and also in the table at the far right; how are these entries related? What is the relation between "170 kilocalorie" and the table at the right about servings? The Principle of Informative Changes has been violated: We need additional marks to help us understand. And the scattered organization of the table itself makes it hard to read. Moreover, is all of this even relevant to the average consumer? Even without knowing in detail what the average consumer really needs to know, we are led to suspect that the Principle of Relevance has been violated. Finally, which part of the display is most important? Where should readers start, and how should they proceed through the display? Without using the Principle of Salience appropriately, the reader is left at sea.

In short, every one of the principles has been violated in this one display!

Do the problems with displays like this mean that we should avoid using graphics in our publications and presentations? Of course not. In fact, in accordance with the Principle of Capacity Limitations, I recommend using combinations of text and illustrations whenever possible. However, I recommend considering all of the principles, both when you design a graphic and then afterward, when you review what you've made. Often it is simply impossible to keep all of the relevant factors in mind when you first conceive and produce a display, and only after you produce a graphic can you then check to ensure that you have not accidentally violated specific principles.

The set of eight principles (also summarized in Appendix 3) will help you create different kinds of displays in the following chapters. Moreover, these principles are elaborated in more detail, as relevant, in the following pages. But if you aren't interested in such information, you can ignore it. My purpose is not to tutor you in facts about perception and cognition, but rather to produce specific recommendations for how to construct good visual displays.

To offer specific recommendations, I draw most heavily on findings about the workings of our perceptual and cognitive systems, which allow the principles and recommendations stated here to be general, to help you to design all types of displays—and give you scope to invent new ones. In addition, I will also rely, to some extent, on the results of studies of how people read charts and graphs. However, the literature on the perception and interpretation of graphs is relatively restricted, and many of the studies used flawed displays as their stimuli. It is difficult to interpret a finding that bar graphs are better than line graphs, say, if the graphs were drawn poorly. Furthermore, the studies vary widely in the tasks they used (asking participants to compare two portions of a display, compare the parts to the whole, extract specific values, or classify trends) and in the measures that were taken (response times, ratings of sharpness of increases, or accuracy). Thus, it is not surprising that the results from different studies sometimes seem contradictory (for reviews of this literature, see Carswell, 1992a; DeSanctis, 1984; Feinberg, 1979; Friel et al., 2001; Jarvenpaa & Dickson, 1988; Macdonald-Ross, 1977; Shah & Hoeffner, 2002; Wainer & Thissen, 1981). I have read these studies with a critical eye and present what I believe to be their essential message.

Finally, this book guides the user to develop effective graphs for normal adults. There is a literature on graph effectiveness in children, which indicates that the quantitative specifications of some of the principles discussed here should be modified for them (see, e.g., Bryant & Somerville, 1986; Curcio, 1987; Mokros & Tinker, 1987). There is also a small literature on graphs for the blind (see, e.g., Aldrich & Parkin, 1987; Lederman & Campbell, 1982, 1983).

The Anatomy of a Graph

Graphs not only let us see that there is more of something in one case than in another, they can also tell us how much more. To understand how a graph provides information, we must look a little more closely at its structure (see also Kosslyn, 1989; B. Winn, 1987).

All graphs, no matter how different individual examples may look, are created from the same components. Typically, they have three primary elements: the framework, the content, and the labels. Thus stripped down, the "graph" in the figure below is reminiscent of the Cartesian graph we met in high school, with its calibrated vertical Y axis and horizontal X axis and origin at 0. In algebra class the content of such a graph was often the curve produced by plotting an equation—a special case of a relationship among numbers.

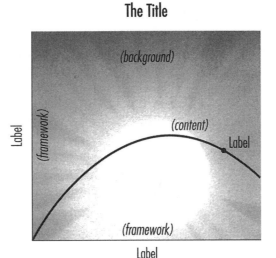

Figure 1.9. The primary elements of a graph—the framework, content, and labels—with the content shown against a background. You are reading a caption, which explains key terms or draws the reader's attention to specific aspects of the graph.

The Framework

The *framework* of the graph sets the stage, indicating what kinds of measurements are being used and what things are being measured. The simplest framework has an L-shape, one leg standing for the amount of a measured substance—*the independent variable*—and the other for the things being measured—the *dependent variable.* The vertical leg, the Y axis, usually stands for the measurements (dollars, barrels of oil, degrees of temperature, etc.), and the horizontal leg, the X axis, for the things being measured (countries, companies, professions, etc.) or the circumstances in which the measurements were taken (years, seasons, elapsed time, etc.).

The Content

The *content* is the lines, bars, point symbols, or other marks that specify particular relations among the things represented by the framework. The positions of content elements typically are plotted as values along the Y axis (e.g., dollars) and are paired with values along the X axis (e.g., seasons). Graphs display information that is associated with different levels of one or more variables. The line graph below illustrates data—fertility rates—about two variables, year and country. The six levels of the year variable are specified along the X axis, and the two levels of the country variable are displayed as separate content lines. A variable that is broken into several content elements is called a *parameter* (the term "stratum" is

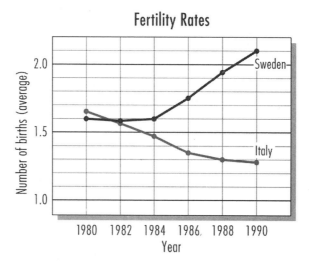

Figure 1.10. This graph has two independent variables, year and country; "Year" has six levels, and "Country" has two.

sometimes used for "parameter"; see, e.g., Lewandowsky & Spence, 1989). Here the parameter is country, with Sweden and Italy being represented by separate lines. In an earlier formulation (Kosslyn, 1989), I used the term "specifier" (also adopted by Carswell, 1992a) instead of "content" because in a sense the entire display is the content. However, to avoid jargon, I will use the term "content" in a more restricted sense here.

Do some visual dimensions specify content better than others? Cleveland and McGill (1984a, 1985, 1986, 1987) suggest that visual dimensions can be ordered in terms of how well people can use them to compare quantitative variations. Based on a mixture of theoretical and empirical findings, they suggest the following ordering of dimensions (from best to worst): position along a common aligned scale, position on identical but nonaligned scales, length, angle/slope, area, volume/density/color saturation (all about the same), hue. Carswell (1992a) reviews the results of 39 experiments and reports some support for this ordering. However, she found minimal differences for position, length, or angle, but area and volume were consistently worse than the other dimensions; these findings suggest not so much an ordering as two categories, with the members of one—area and volume—being inferior to members of the other. In addition, Carswell found that the ordering of visual dimensions depends in large part on the type of task. This ordering fares better at predicting performance when participants must focus on one portion of the graph (as when specific point values had to be extracted) than on tasks where different portions must be integrated or compared; indeed, the predictions of Cleveland and McGill's ordering were actually contradicted when participants had to synthesize information from different portions of a display (e.g., when determining whether the variability of the data points was large). Also, their claims did not fare well when participants had to recall graphed information.

Spence and Lewandowsky (1991) report findings that suggest that Cleveland and McGill's ordering reflects only the initial phases of registering a display (see also Cleveland, 1985; Cleveland & McGill, 1987). Spence and Lewandowsky asked participants to use different types of displays to make local and global comparisons within a specified period of time (under a "deadline"); the ordering of visual dimensions predicted the results best when the deadline was very short. In addition, Simkin and Hastie (1987) report that the relative efficacy of different visual dimensions for conveying information about quantity depends in part on the specific task. Moreover, Peterson and Schramm (1954) asked aviators to use graphs that specified information with angles or position on aligned scales, and found that the airmen tended to be better with angles. These findings are interesting because other types of participants show the opposite pattern. Carswell (1992a) argues that because aviators are trained to read circular displays, these results may provide evidence that training affects the relative ease of using specific types of content. Indeed, DeSanctis and Jarvenpaa (1985) showed that with practice business students were better able to use graphs to make financial forecasts—but that practice improved their performance only with a "standard" horizontal bar graph format and not one with scales that had different maximal amounts and nonround value labels. It is clear that the accuracy of simple judgments of amount is not the only factor you should consider when deciding how to display data. The recommendations offered in this book are based on the wider set of principles summarized in above and in appendix 3.

The Labels

Each leg of the framework bears a label naming a dependent variable (the type of measurement being made) or an independent variable (the entity to which the measurement applies). In "Fertility Rates," the dependent variable is the birth rate ("Number of births [average]"), calibrated along the Y axis, and the independent variable along the X axis is "Year." Other labels indicate values along the measurement scale (here, specific average number of births—1.0, 1.5, 2.0), and the particular entities to which the measurements apply (the years 1980, 1982, 1984 . . .). The title of the graph is itself a kind of label. If the content has a parameter, this independent variable (in this example, country) and its levels also typically are labeled. If the names of the levels—here, Sweden and Italy—are sufficiently distinct, the name of the parameter variable itself ("Country") can be omitted, as it is in this case.

Optional Components

Graphs also may include a number of optional components. For example, some graphs may include *inner grid lines*, as does "Fertility Rates." These lines stretch across the framework at regular intervals, either horizontally, vertically, or in both directions. These lines carry no meaning in and of themselves; they simply make it easier to trace along from a point on one leg of the framework to a content element and from there to the corresponding point on the other leg of the framework. Grid lines sometimes can be useful to the reader because, as noted earlier, information about location and shape are not always precisely combined in the brain.

Some graphs also include a *background*. For example, a graph about infant development might be superimposed over a picture of a baby. The background serves no essential role in communicating the particular information in the display; if it were eliminated, the display would still communicate the relevant information. Backgrounds can convey the general topic, and can sometimes make a display flashier and more eye-catching, but we will be concerned about backgrounds not so much because of what they can do *for* a display, but because of what they can do *to* a display—make it difficult to understand.

Graphs sometimes include a *caption*. A caption is a comment on the display, a short description that explains key terms or directs the reader's attention to specific features of the display. Captions are common when displays appear in textbooks but are seldom seen in magazines, trade books, or presentations. In some cases, titles and captions are combined, creating a long, discursive title.

Putting It All Together

This breakdown of a graph into components will help us in two ways. First, from a practical point of view, it allows us to avoid redundancy. Recognizing that a framework, labels, and title are common to most graph types—bar, line, layer, scatterplot—allows the recommendations for producing those elements to be grouped, as they are in chapter 3.

Second, and perhaps more important, this way of looking at graphs helps us to see how graphs work well. To obtain a precise value in a line graph, the reader not only must register the components of the graph but also must relate a specific point on a content element (a line) to the corresponding locations on the X and Y axes. To be effective, therefore, not only must a graph include the essential elements, but also these components must have the proper relations. A graph is more than the sum of its parts; the components must be organized in a way that facilitates our seeing the relations among them. Much of the material in the following chapters addresses this concern.

And this brings me to a cautionary note. A graph's components can be arranged in so many ways that it's impossible to anticipate the effects—positive or negative—of every possible way to plot a particular data set. Fortunately, the psychological principles in this book will always apply to all types of displays. Thus, even if none of the specific recommendations that I offer applies to a decision you must make in choosing or designing a display, the principles will guide you.

In particular, keep in mind at all times that a graph is intended to make a specific point, and its visual appearance should convey that point. Many of the psychological principles underscore the importance of context, and in some circumstances may contradict a specific recommendation. An example: If your point is to show that a market has become badly fragmented, including many tiny slices in the same pie graph would be a *good* idea—even if each slice cannot be easily discriminated or interpreted. The recommendations are not hard-and-fast rules but handy guidelines. They usually will be appropriate, but always keep in mind the point you are trying to make and let the clarity of your message be the ultimate arbiter.

The Next Step

Variations in the different components produce a wide range of graph types—line, bar, pie, scatterplot, and more. The next step is to pick one. Chapter 2 presents guidelines to help you decide which type of graph to use in a specific situation. My recommendations for choosing the appropriate type of graph rest on two factors: the particular kind of data you have to present, and the particular types of questions you want the reader to be able to answer. In the course of considering the options, we'll explore the principles in more detail.

Choosing a
Graph Format

2

T HE FIRST STEP in making a good graph is deciding what kind of graph to use. If you choose the wrong type for the task at hand, the graph will not communicate effectively—no matter how well it is designed. In order to move toward a choice, you should consider several fundamental questions, not the least of which is whether you should use a visual display at all. Gnanadesikan (1980; cited in Wainer & Thissen, 1981, p. 196) introduced a number of criteria to consider when choosing a display, such as its potential for internal comparisons and aid in focusing attention. Many of his criteria are similar to those offered here. However, Gnanadesikan appears to have neglected a central point I wish to stress: The usefulness of a graph can be evaluated only in the context of the type of data, the questions the designer wants the readers to answer, and the nature of the audience.

To Graph or Not to Graph?

The following recommendations will help you to decide whether you should use a graph, or would be better off with a table or discursive presentation.

Use a graph to illustrate relative amounts.

Use a graph only if the point is to illustrate relations among measurements. As we have seen, graphs use variations along a visual dimension, such as the height of a bar or line over a specific point of the framework, to convey quantitative information. Our perceptual systems allow us quickly to detect differences among heights or the slopes of lines. However, they do not allow us to register absolute heights or slopes very well, and thus graphs are not well suited for conveying specific absolute values (Principle of Capacity Limitations). The problem is this: In order to derive a specific value, readers must look from the content element to the scale, and we are not very good at registering precise spatial relations of objects that are not physically connected (Principle of Compatibility, the specific aspect of spatial imprecision).

The reason why we must pay attention to register spatial relations is that separate brain systems process information about shape and location. Input from the eyes first is processed on the surface of the back of the brain, and then information flows forward along two major pathways. Spatial properties, such as location and size, are processed in the top rear portions of the brain, whereas object properties, such as shape, color, and texture, are processed in parts of the brain that lie under the temples. It is only relatively late in the processing that information about spatial properties and object properties come together in the brain. To register spatial relations among shapes precisely, one must pay attention carefully to two parts of an object or scene. It is often difficult to pay attention to a content element and the scale at the same time and still see both clearly.

We rarely are aware of the precise spatial relations between parts of a display. This limitation underlies one facet of the Principle of Compatibility: The strong suit of graphs is the illustration of quantitative relations, and they are not appropriate if you want to convey only precise values. If this is your goal, use a table.

There are a remarkably large number of studies that were designed to compare the relative efficacies of graphs and tables of numbers (for reviews, see Casali & Gaylin, 1988; DeSanctis, 1984; Jarvenpaa & Dickson, 1988; MacGregor & Slovic, 1986; Meyer, 2000; Meyer, Shinar, & Leiser, 1997). The general finding is that graphs are better than tables of numbers only for specific purposes. For example, Washburne (1927) found that tables of numbers are better for recall of specific amounts, whereas bar graphs are better for complex comparisons, and line graphs for trends (as cited in Umanath & Scamell, 1988). Umanath and Scamell (1988) compared a table to a bar graph and found that the graph was better for recall of rank order and pattern information, whereas the table was as good as the graph for specific point values; similarly, Spence and Lewandowsky (1991) found that both pie and bar graphs were superior to a table when relative values had to be compared—even if only a few numbers were involved (counter to the suggestion of Tufte, 1983, to use a table for small amounts of data; see also Carswell & Ramzy, 1997).

However, although Benbasat and Schroeder (1977, cited in Lucas, 1981, p. 758) found that graphics displays were better than tables in an inventory management task, others (e.g., Lucas, 1981) found no clear difference. And yet, Moriarity (1979), Stock and Watson (1984), Nawrocki (1973), and Schwartz (1984) all found that participants performed better with graphics displays than with tables in decision-making contexts. In addition, Casner and Larkin (1989) showed that a graphical format was superior to a table when participants had to use data to make airline reservations. The graphs they used were designed to reduce the effort of using specific mental processes and to reduce the amount of search necessary; these considerations have been incorporated into the principles and recommendations offered in this book.

Perhaps the most surprising results in this literature were reported by Meyer et al. (1997). They presented line graphs, bar graphs, or tables and asked participants to read precise values, compare pairs of values, identify trends, or read maximum values. In all cases, participants were faster with tables and were at least as accurate with tables as with graphs. The participants could identify trends in line graphs more quickly than they could identify trends in bar graphs, but that was the only such difference these researchers found.

On examination of the sample stimuli provided by Meyer et al. (1997, p. 273), I note three possible reasons for these findings. First, all graphs included keys, not direct labeling; in contrast, the rows of the tables were directly labeled at their left. Thus, the graphs required finding the key and then locating the corresponding line or bar—which required time. Indeed, Carpenter and Shah (1998) found that readers continually look back and forth from labels when examining graphs (even when the lines were directly labeled). Second, the keys were not easy to use: Not only were they located at the bottom, not in the customary upper right-hand corner, but also only a very short segment of the corresponding line was presented. Third, the data appear to be essentially random. Thus, there is no recognizable pattern in the graphs. As I will argue repeatedly, drawing primarily on the Principles of Compatibility and Capacity Limitations, graphs are particularly useful when the visual pattern they convey maps directly onto a familiar interpretation, such as a linear trend or cross-over interaction (and Meyer et al., 1997, p. 285, also recognized that this characteristic might help to explain their results).

A striking aspect of Meyer et al.'s (1997) article is that—in spite of their finding that tables are superior—they themselves nevertheless present the results using graphs! The authors acknowledge this "apparent contradiction" (p. 285) and justify it by noting that graphs are likely to be noticed by someone who is merely skimming a paper and graphs facilitate the display of interactions (which is what they mainly used their graphs to show).

Subsequently, Meyer, Shamo, and Gopher (1999) showed that graphs are in fact more effective than tables when the data are not random, especially when the task requires making use of the structure in the data. With respect to the way graphs experiments are conducted versus the way graphs are actually used in practice, Meyer et al. suggest that

> the actual use of displays differs from most experiments in at least one respect: Outside the laboratory the displayed information usually has meaning (i.e., is nonrandom) and therefore holds some kind of structure. In contrast, most laboratory experiments that compare graphic and tabular displays use random data or data that have no systematic structure. (p. 571)

To investigate the force of this observation, these researchers asked participants to compare pairs of data points or to identify the direction of trends in the data. They compared performance not only when the participants were shown tables versus line graphs, but also when they were told nothing about the data versus were given general or specific information about the structure in the data (i.e., that they were produced on the basis of sine functions). As expected, Meyer et al. (1999) found that graphs conveyed structured data better than did tables, and were particularly useful when trends had to be identified. Moreover, being armed with knowledge about how the data were created also helped the participants use graphs—especially when graphs were being used to extrapolate from a data set to predict the future.

These findings led Meyer et al. (1999) to extend the idea of task/display compatibility, which stresses that a display must make explicit and accessible the information needed to perform a task (see also Wickens & Carswell, 1995). Instead, they advocate considering the interaction between task, display, *and* data. Their view is that people actively search for structure in data and that graphs are useful to the extent that such structure is available (see also Coll, Coll, & Thakur, 1994; Oron-Gilad, Meyer, & Gopher, 2001). Consistent with this view, Umanath and Vessey (1994) found that graphs were better than tables when the participants had to use the data to predict future events (in this case, the probability that a company would go bankrupt).

Finally, for some tasks, graphs are particularly useful only if they distinguish between relatively important versus unimportant information, varying visual salience so that readers can immediately sort the grains of gold from the sand. For example, Sanfey and Hastie (1998) gave participants information about a group of runners (their age, total training miles, fastest race, and level of motivation) and asked the participants to predict the runners' finishing times. In this task, presenting the information in a table or text led participants to use the data more effectively than did presenting the information in bar graphs. Sanfey and Hastie infer that the textual materials led participants to order the cues more appropriately in terms of their relative importance. "Probably the best summary statement is that bar graphs encourage subjects to weight each of the cues approximately equally, at least in comparison with the other formats, but do not induce an advantageous differential weighting of the cues" (p. 103). This finding does not indicate, however, that all graphs would be equally ineffective. Indeed, MacGregor and Slovic (1986) found that when a display emphasized the relevant aspects of the data in this decision-making task, it was more effective than a table.

In short, although graphs are not always superior to tables, the bottom line—as I read it—is that graphs are more effective than tables when relations among values are critical, especially when systematic, important relations are to be conveyed (see also Jarvenpaa & Dickson, 1988; B. Winn, 1987). Note, however, that the effectiveness of tables—like graphs—depends on how they are designed (see Ehrenberg, 1975). If you want to convey both relations among data and absolute values, consider putting a few numbers in critical places on the graph.

Specify the subject.

What do you want your readers to know after examining the display? What information will they need? The Principle of Relevance leads us from the outset to decide what will be relevant in the display.

One useful way to decide what to put in a display is to formulate a precise title. "Plant Productivity, 1940–1990" would lead you to include information different from that in "Number of Units Produced in the United States and Japan." In the first case, you would supply data for each of the different years; in the second, you would simply average the annual production figures to show totals for each country. A display titled "Change in Productivity in the United States and Japan, 1940–1990" would include data by both year and country.

Present the data needed for a specific purpose.

A graph is a device to help people answer specific questions. You must decide what specific questions the readers will want to address in the context of the graph, and precisely what information readers will need to know to be able to answer those questions from the display, and organize the data accordingly.

This recommendation grows out of rules that govern our verbal communication. In particular, the linguist Paul Grice (1975) formulated rules that ensure smooth discourse between speakers. One of his rules is that people expect a question to be answered with the appropriate amount of information, no more and no less. This rule can be extended to visual displays, as stated by our Principle of Relevance. Readers use displays to answer particular questions, and they expect to be told as much information as is necessary to answer those questions in the context in which the graph appears—but no more.

You should not include any more or less information than is needed to make your point. The Principle of Relevance is violated when, as often happens, the information in a graph is broken into categories that are extraneous to the purpose of the display. For example, if you want the readers of your annual report to know whether overall sales are increasing at the same rates in different parts of the country, plotting the data separately for each of your product lines would only be an obstacle; readers would mentally have to average over lots of lines or bars to obtain the information they needed. Similarly, if variation over years is what is important, average over individual months and do not present the individual months in the graph. On the other hand, it is important to provide enough detail to convey the message. If you want the readers to know about differences in sales of commercial and personal equipment, it would be a fatal error not to include separate bars or lines for each type—there is no way of mentally breaking down a single point or bar into its constituent parts.

Use concepts and display formats that are familiar to the audience.

The display must be designed to communicate to a particular audience. The concepts used must be familiar to that audience, and the display format should be comprehensible to it. This advice stems from the Principle of Appropriate Knowledge: Readers can know how to interpret a display only if they have already stored the necessary background in memory. Plotting first or second derivatives and labeling an axis with those terms excludes people who have not studied calculus. Know your audience, and present your information accordingly.

For example, an experienced graph reader seeing a display with one rising line and one falling line forming an X pattern can immediately recognize that pattern as describing one kind of interaction, that is, a situation in which the effect of the value of one independent variable depends on the value of another. Perhaps goats weigh more than sheep in the summer, but vice versa in the winter. The graph of this situation is an interaction, because the value of one variable (goats versus sheep) depends on the value of the other (whether it is winter or summer). If season is graphed on the X axis and weight on the Y axis, and goats and sheep are separate lines, the line for goats will cross the line for sheep (provided that the animals weigh about the same amount on average), forming an X pattern. By taking time to analyze the picture presented by the graph, any reader will eventually come to the correct conclusion, but only readers who have seen and analyzed a number of such graphs will be able to associate the X pattern itself with a meaning stored in memory and thereafter know the significance of the pattern immediately.

To understand something, we must grasp its implications and notice its relations to other things. Visual displays often are more like stories than like pictures of objects; although they have components that must be identified, it is the relations among the components that convey the specific information. This aspect of understanding in part requires recalling the significance of particular patterns, and may require reasoning based on the rules of how a display format works.

Pinker (1990, p. 109) identifies 12 visual patterns of lines that signal distinct quantitative trends: flat (unchanging), steep (increasing rapidly), inverted U (quadratic), U (quadratic), jagged (random), undulating (fluctuating regularly), straight (linear), S-shaped (cubic), rectilinear (abruptly changing), not flat (variable X affects variable Y), parallel (variable Z has additive effects on those of variable X), and converging (variables X and Z interact). These patterns can be more finely distinguished; for example, there are four distinct ways in which lines can converge. In short, an expert graph reader has a remarkable amount of knowledge about interpretation of patterns of lines.

Pinker argues that graphs are useful in large part because experienced readers develop specific "graph schemata" in memory, which allow them to identify meaningful patterns. Simkin and Hastie (1987) develop this idea and present empirical support for it (see also Wainer & Francolini, 1980). Moreover, W. D. Winn (1983), in studies of eye-movement patterns, found that people quickly fixate on meaningful shapes when information is presented graphically.

Even if the data do not produce easily identifiable patterns, familiar formats are easy to read because the reader can effortlessly carry out a sequence of steps. A demanding task is one that entails many steps and requires one to attend to each of them. As one becomes familiar with a task, it becomes less demanding because one organizes the steps into larger units and no longer needs to attend closely to the individual steps; with practice, we develop "automatic" processes. For example, when first learning to drive, the student must pay attention to every detail; shifting gears with a manual transmission requires pressing in the clutch, letting up on the gas, pulling the stick, pressing down on the gas, and slowly letting up the clutch. With practice, the entire process becomes organized as a single operation, and one no longer needs to pay attention to the details (see Anderson, 1981, 2004; Osherson, Kosslyn, & Hollerbach, 1990; Smith & Kosslyn, 2006). Indeed, these kinds of habits might even use a different brain circuit, which directly connects inputs to responses (see Mishkin & Appenzeller, 1987).

In this vein, it is worth noting that Phillips et al. (1990) found that the precise nature of symbols matters more to inexperienced readers (of maps, in their study) than to experienced readers. Moriarity (1979), Stock and Watson (1984), and MacGregor and Slovic (1986) show that relatively exotic displays can be used effectively if readers learn how to interpret them. (See also Ehrenberg, 1975; Jarvenpaa & Dickson, 1988; Simkin & Hastie, 1987; Vernon, 1946; but also see Macdonald-Ross, 1977, for a critique.)

Selecting the Right Format for the Job

After you have decided what numbers to present and have thought about your audience, you are ready to consider particular formats. Your choice of graph format will depend on the kind of information being conveyed and on your specific purpose in presenting the display. In this chapter, I discuss the most familiar types of displays, which are standard in most computer graphics programs. I discuss here only those types of graphs that are commonly used in newspapers, magazines, and other nonspecialized media; for more sophisticated and detailed taxonomies and discussions of special-purpose displays, see Bertin (1983), Chambers, Cleveland, Kleiner, and Tukey (1983), Cleveland (1985), Feinberg (1979), Macdonald-Ross (1977), Tufte (1983, 1990), Tukey (1972, 1977), Wainer (1979), and Wainer and Thissen (1981).

Keep in mind, however, that although some of these exotic displays may be better than conventional displays in laboratory tasks, they can fall short in more natural settings (e.g., Goldsmith & Schvaneveldt, 1985, found a "star" display to be better than a bar display in the lab, but Peterson, Banks, & Gertman, 1982, found no difference in the context of a nuclear power plant). My own view is that the conventional displays have survived a kind of Darwinian winnowing process, and the mere fact that they continue to be employed over so many years (since their invention by Playfair, 1786, in some cases) is itself evidence of their utility.

Graphs for Percentage and Proportion Data

How is an operating budget divided among salaries, energy costs, benefits, and so on? What is the contribution to total profits of each division? What percentage of the national Republican vote came from each region of the country? These are questions about how a whole is divided into parts; numbers of this sort must add up to a fixed total, 100% for percentages, 1.0 for proportions.

Percentage and proportion data can be presented using a variety of formats, including bar graphs and line graphs. However, if you want to emphasize that the components sum to a single whole, then you should choose a pie or divided-bar format. These formats make clear not only the relations among the various components but also their relation to the whole. (If you want to display components of a whole that do not add up to a constant amount, see the discussion on stacked-bar graphs and layer graphs below.)

Pie graphs are circular, with the proportions of the whole indicated by the sizes of wedges; divided-bar graphs are rectangular, the proportions of the whole indicated by the sizes of internal segments. Pie graphs do not have a labeled scale, but rather represent amount by variations in the size of content elements; divided-bar graphs can have a labeled scale. In both formats, the framework is implicit in the external boundary of the content. For these kinds of displays, the reader can easily extract precise amounts only if they are specified by separate labels.

When to Use a Pie Graph

The most common way to display how a single whole is divided into parts is to use a pie graph. However, there has long been a controversy about just how well human beings can determine the area of pie wedges, which might challenge the idea that we should use pie graphs at all. Eells (1926) reported the first study of the ability to read quantity from area and found that people read the relative proportions of pie wedges more accurately than they do segments of divided-bar graphs, particularly when more components are included (but see Von Huhn, 1927, for a critique). Croxton and Stryker (1927) report similar findings but found that divided-bar graphs were better than pie graphs in some circumstances (when there were only two parts, and they did not display a 50:50 or 75:25 relation).

Macdonald-Ross (1977) recommends bars over pie graphs because people systematically underestimate area. However, Macdonald-Ross's recommendation was based on judgments of the relative size of entire forms, not their components. Similarly, although Cleveland and McGill (1984a) found that people cannot judge area well from pie graphs, they asked participants to judge the percentage a smaller element was of a larger one— which is not what we usually do when reading graphs. Simkin and Hastie (1987) found that participants could compare two quantities (and determine what percentage of the larger is represented by the smaller) most accurately for bars, then divided bars, and then pie segments; in contrast, when asked to determine the percentage of the whole, segments of pies and bars were judged more accurately than were segments of a divided bar. (However, Simkin & Hastie did not include a scale on the divided-bar graphs, which probably would have facilitated performance for these displays.)

Spence (1990) asked participants to judge the relative proportions two parts are of a whole (rather than direct size comparisons) and did not find systematic underestimation when only one dimension (e.g., wedge size, height) was varied. Indeed, Spence found that participants could compare the relative proportions of pairs of bars or pairs of pie elements equally well when only a single dimension was varied. However, Spence found that pies were not quite as good as bars when viewers had to switch back and forth between different formats. But Spence and Lewandowsky (1991) found that a pie graph can actually have a slight advantage over a bar graph if the judgment requires complex comparisons of components.

Perhaps of greatest relevance, Hollands and Spence (1998) found that pie graphs represent proportions better than do bar graphs, and also found that the time to make judgments of proportions increased with each additional bar in a bar graph—but not with each additional wedge in a pie graph. Hollands and Spence provide evidence that participants performed this task by mentally summing the bars in a bar graph and comparing individual bars to the total, which is not necessary with a pie graph (and hence for this task, pie graphs respect the Principle of Capacity Limitations—the aspect concerning processing limits—and bars do not).

Thus, the task and type of display must be jointly considered when we try to decide which type of display is best for a specific purpose. Indeed, Simkin and Hastie (1987) provided strong support for the view that graphs should directly present the requisite information; they showed that for some purposes some types of displays require additional, time-consuming mental operations (e.g., to compute a trend on the basis of a set of bars). In addition, in careful psychophysical studies, Hollands and Spence (2001) compared the efficacy of pie graphs and divided bars for different purposes. In particular, they examined cases in which participants had to compare corresponding portions of different graphs, which could vary in overall size. On the basis of their results, they conclude that if "a series of graphs depicting proportions or percentages consists of equal-size wholes, divided bars are preferred. If, however, the wholes are of unequal size, a series of pies is the better choice" (p. 430). Varying the overall size of a pie does not disrupt the relative angles of the wedges, but varying the size of a divided bar does disrupt the actual heights of segments within bar. Similarly, in later work Hollands (2003) showed that pies are better than stacked bar graphs when the overall size varies.

Clearly, we need to temper Edward Tufte's (1983) assertion (based solely on his intuition, as far as I can tell) that "the only design worse than a pie chart is several of them" (p. 178). In some situations, this opinion is no doubt justified, but we should not make such a sweeping generalization about the value of any type of display, independent of the type of data to be displayed and the purposes to which the display will be put.

The recommendations I offer in this book are consistent with the major themes of this literature, but I have given more weight to the results from studies that used more realistic tasks and displays.

Use a pie to convey approximate relative amounts.

A pie graph effectively conveys general information about proportions of a whole. However, if the reader is supposed to obtain relatively precise amounts, such as the percentage of one part, it is better not to use a pie; this information cannot be easily obtained because it is difficult visually for us to measure the angles, chords, or areas of wedges precisely. It is very difficult, for example, to use the **don't** graph to see that there were thirteen percent more car and mobile home loans than bank and finance loans: The Principle of Compatibility (specifically, the aspect concerning perceptual distortion) is at work. Some dimensions are systematically distorted by our visual systems; specifically, our visual impression of area is less than what is actually present, and the underestimation is more severe for larger regions. This is a problem with pie graphs because research has shown that about one-fourth of graph readers apparently focus on relative areas of wedges when they read such graphs—which means that they will systematically underestimate the sizes of larger wedges.

For example, Eells (1926) found that most (fifty-one percent) of his participants read relative areas of pie graphs by looking at the arcs, twenty-five percent by estimating area, twenty-three percent by estimating the angle at the center, and one percent by estimating the length of chords (i.e., straight

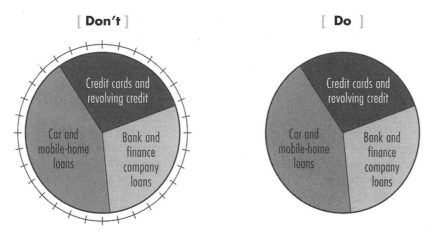

[**Don't**]   [**Do**]

Figure 2.1. Do not use a scale with pie graphs; the reader will have to struggle to count the number of ticks. In this graph, credit cards and revolving credit are the topic of interest, and hence this information is most salient.

lines connecting the ends of the lines defining the wedges). Unfortunately, not only do our visual systems distort area, but also they fail to register angle with great precision. Relatively small acute angles tend to repulse each other (see Howard, 1982; see also Schiano & Tversky, 1992). In addition, angles that are symmetrical around the horizontal axis are seen as larger than angles of the same size that are symmetrical around the vertical axis (Maclean & Stacey, 1971).

The systematic distortion of area is captured by "Stevens's Power Law," which states that the psychological impression is a function of the actual physical magnitude raised to an exponent (and multiplied by a scaling constant). To be precise, the perceived area is usually equal to the actual area raised to an exponent of about 0.8, times a scaling constant (see also Brinton, 1916; Macdonald-Ross, 1977); similarly, the perceived brightness is usually equal to the actual brightness raised to an exponent of about 0.7, times a scaling constant. In contrast, relative line length is perceived almost perfectly, provided that the lines are oriented the same way (see Baird, 1970; Stevens, 1974, 1975). (Note, incidentally, that pain has the opposite relation: increasingly less electricity—used to deliver an electric shock—is required to induce an increment of discomfort as the general level of electricity increases.)

However, for our purposes it is important to note that Teghtsoonian (1965) found that the exponent for area varied depending on exactly what the participants were instructed to do: The exponent was close to 1 (representing veridical perception) if participants were asked to judge "real," as opposed to "apparent," area. Spence (1990) found that exponents were close to 1 when participants compared the relative proportions of two pie wedges, vertical lines, horizontal lines, disks (pies seen from above and to the front), bars, boxes, and cylinders, provided that only one dimension was varied at a time. Spence did not explicitly ask participants to judge the "real" extents, but suggests that the use of graphical elements alone might have led the participants to make judgments of real, not apparent, size (see also Meihoefer, 1969, 1973).

Although the distortion can be overcome, this may require motivated readers who are willing to reason their way past initial visual impressions. In general, such impressions will be distorted: The larger two regions are, the greater the percentage of difference will have to be in order to be seen as equivalent to a difference between two smaller regions. You cannot correct for this error by distorting the graph because individuals differ in the degree of error, and the degree of error depends on the precise material being compared (see also Cleveland, Harris, & McGill, 1982).

Although our perception of the areas of segments of divided-bar graphs is not systematically distorted, Eells (1926) found that almost three times as many people preferred pies to divided bars. (These data are admittedly a bit long in the tooth, but there is no reason to think that people have changed in the interim; indeed, pie graphs are probably even more common today than in 1926.)

In general, one way to convey precise amounts while still depicting relative proportions with pie graphs is simply to label the wedges, either by putting the numbers directly in the corresponding wedges (if space permits) or next to them. This converts the pie into a hybrid display, part pie and part table. Such redundancy can be a good thing, providing the reader with two ways to get the message—if precise values are part of what you want to convey.

Use an exploded pie to emphasize a small proportion of parts.

An exploded pie display is constructed by displacing the important slice or slices, as if a wedge of pizza had been pulled out from the pie. If the reader is supposed only to make approximate visual comparisons, the pie format provides a particularly good way to draw attention to a small percentage of the total number of components. Look at the **do** graph; it is immediately obvious that credit cards and revolving credit are the most important aspects of the display. In contrast, in the **don't** version it is not clear what the reader should focus on.

Exploded pie charts can be effective because of how our neurons (nerve cells) work. Neurons in the visual system can be thought of as "difference detectors"; they respond most strongly to a change in stimulation. We are led to focus first on the features of a display that are brighter or darker, are in motion, or are in some other way different from the surrounding parts of the display. This is an aspect of the Principle of Salience, which states that

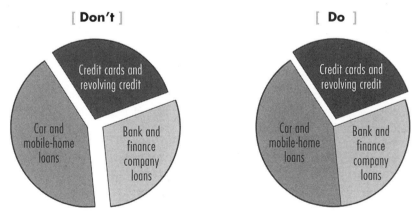

Figure 2.2. Use an exploded pie only when a distinct contour can be disrupted by exploding the wedges; the two intact pieces of the pie on the right define a contour, which makes the exploded wedge stand out.

attention is drawn to large perceptible differences (for examples of the role of variations in salience in graphical perception, see Brown, 1985; De Soete & De Corte, 1985). An exploded slice or two that disrupts the outer contour of the framework will be noticed immediately. This format should be used only if a clearly defined contour remains after the key parts are exploded (two wedges exploding are too many in the example).

If proportions vary greatly, do not use multiple pies to compare corresponding parts.

Compare the two pies in the **don't** version. How did Peugeot and Renault do in the European Community (EC) in general (left) compared with Italy in particular (right)? This is difficult to fathom, in part because the relevant wedges are in different locations in the two pies—a shift that cannot be avoided when the proportions vary widely in the two pies. Now compare the two **do** pie graphs; it is clear that there is a rough correspondence between the revenues per region (left) and employees per region (right). It is easy to compare multiple pies when the wedges are in roughly the same positions in each, but when the corresponding parts are in different locations, the reader is forced to search for them one at a time. Remember the Principle of Capacity Limitations; our brains have limited processing capacity, and if forced to strain, many readers will simply give up.

Therefore, if the proportions are very different, use the pie format only if the point is to show that there is a major change over the levels of the parameter (with a different pie corresponding to each different level). This recommendation is rooted in the Principle of Informative Changes, which leads readers to expect any change in a pattern to mean something (this is a variant of Shannon's [1948] classic formulation of the nature of "information"). When things stay the same, there is no new information; when something changes, there is—or should be—new information. Readers expect every noticeable change in the appearance of parts of a display to mean something; if it does not, the change is simply a distraction. We are not passive viewers, but rather build up expectations about the world as we see and actively seek specific information—which requires effort (Gregory, 1970; Kosslyn, 1994; Neisser, 1967, 1976).

In short, if the proportions are very different across levels, and the reader is supposed to compare specific components, a bar graph is the preferable format.

Car Markets

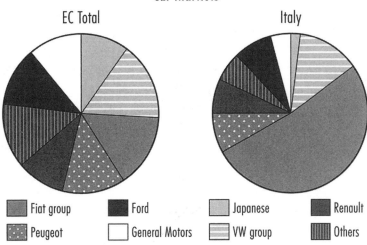

Revenues and Employees per Region of Electrotechnical Corporation

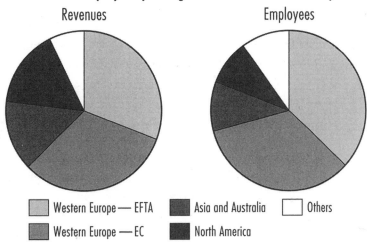

Figure 2.3. Corresponding wedges are hard to compare if, as in "Car Markets," they are not in corresponding positions.

When to Use a Divided-Bar Graph

As noted, in divided bar graphs the length of each segment within a bar represents the proportion of one component of the whole.

Use a divided bar to convey accurate impressions of parts of a whole.

Unlike area, represented by a pie's wedges, distance along a single extent is not distorted by the visual system; the reader can gain an accurate impression of amounts by noticing the height of a segment. Moreover, the reader can extract precise amounts in this format if direct labels (specifying the values numerically) are included in the display. But for making comparisons among values, direct labels lose the power of a visual display; they require adding and subtracting rather than comparison of amount or extent. To convey a sense of the amount of each segment, use a scale with divided-bar graphs. Cleveland and McGill (1984a) found that readers do not estimate the lengths of segments of divided bar graphs very well. Thus, it is critical that a scale be included if relatively precise impressions are required.

However, a disadvantage of using a scale with divided-bar graphs is that readers must realize that the absolute height of a segment above the X axis

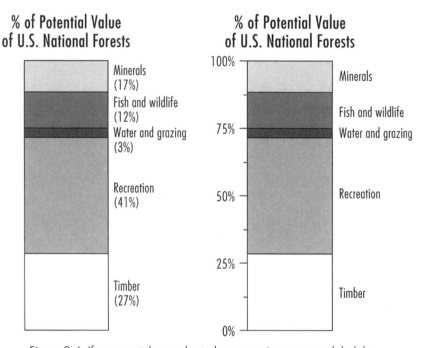

Figure 2.4. If you want the reader to know precise amounts, label the segments directly. A scale is acceptable if you want to give the reader an impression of the actual amounts of each segment, not to convey precise values.

usually is misleading; it is the vertical length of an individual segment (the distance between its bottom and top) that is important. In the example, the relative value of recreation is not sixty-eight percent, but forty-one percent. Readers must count and subtract numbers to get exact percentages—they cannot simply read off numbers. If the subtraction is too difficult, the process will violate the Principle of Capacity Limitations (Kahneman, 1973). A good graph can be read at first glance; it is not a puzzle to be solved.

Graphs for Quantitative and Rank-Order Data

How many units were produced, how much money was spent, how many people were involved? To answer this kind of question, especially when you are not interested in the relation between individual amounts and the sum total, the following formats are most useful:

- Visual tables
- Line and bar graphs
- Side-by-side graphs
- Step graphs
- Scatterplots

These formats can be used to convey rank-order or interval information (numbers on a continuous scale, e.g., money, weight, or temperature). They also can be used for percentage and proportional data if you want to emphasize comparisons among relative amounts; they are less useful for communicating how the amounts add up to a fixed whole.

When to Use a Visual Table

Visual tables are pared-down graphs, which convey information by properties of the content alone; objects are drawn so that their sizes or numbers vary in accordance with the amount being represented. Some visual tables use pictures of actual objects, and some use abstract content such as bars. A special case is the *isotype*, in which bars are created by repeating small pictures, each of which corresponds to a unit, a particular amount of the measured substance (Macdonald-Ross, 1977; Neurath, 1974). Isotypes are particularly interesting because different pictures can indicate component parts of the measured entity; for example, one could specify the number of male and female employees in the different divisions of a company by showing rows of pictures of a man or woman next to a label for each division;

each picture would stand for an increment of, say, 100 employees, with more pictures (forming a longer row) indicating more of that type of employee in that division. In this case, then, the bar would label itself; a reader would not need to consult either a key or written label.

A visual table does not have a framework and hence has no scales; the information is conveyed solely by the relative sizes of regions, exploiting the Principle of Compatibility (the "more is more" rule).

Use a visual table to convey impressions of relative amounts.

If the reader is expected to gain only a general sense of the relative differences among amounts, then a visual table is appropriate. These displays eliminate unnecessary material, in accordance with the Principle of Relevance. A set of water bottles, varying in height (keeping width constant), could be enough to illustrate the increase in consumption over time. But because visual tables do not have scales, and because the visual system tends to distort area, visual tables cannot be used to depict trends with precision.

However, you can label each content element with its amount (in the same way that you can label the sizes of individual pie wedges), allowing the elements to convey precise information. For example, if the sizes of balloons (presumably filled with hot air) indicate the number of speeches delivered by each of several politicians, you could put the name of the politician on each balloon and have a tag with the actual number of speeches attached to the string. But be careful: Not only will readers fail to register relative areas accurately, but also this practice can easily result in a cluttered display—relegating the content elements to the role of mere decorations. If you want to convey precise numbers, consider simply using a tabular format instead.

When to Use a Line Graph or a Bar Graph

Line and bar graphs are the most common formats used to display quantitative data, both in technical and popular media (Zacks, Levy, Tversky, & Schiano, 2001). Both types of graphs are often used to display changes over time. Bar graphs have the standard L-shaped framework and use bars as content elements; the heights of the bars specify discrete amounts. Line graphs are just like bar graphs, except that a line is drawn instead of bars; the height of the line at each point indicates the amount, and changes in the height of the line indicate changes in amounts.

As in any display format, the choice between line graphs and bar graphs depends in part on the nature of the entities being measured and in part on the purpose of the graph—specifically, the question you want readers to answer with the display. Many studies have been conducted to compare how well readers can use the two formats. For example, Shah, Hegarty, and Mayer (1999) showed that the same data plotted in a bar graph convey a different impression than when they are plotted in a line graph. In particular, Shah et al. showed that differences in trends can be depicted much more clearly in line graphs, which also reveal interactions that are buried in the corresponding bar graphs. In addition, Shah et al. showed that the way the information in the two types of displays is organized also influences how well they convey specific information. Shah et al. manipulated the perceptual organization of the material in line graphs by varying which dots were connected to form lines and manipulated the perceptual organization of bar graphs by varying which bars were placed nearby to form visual "chunk" (perceptual unit). As expected, the type of organization had a substantial impact on how the participants described the information being conveyed by a graph: Participants were more likely to make comparisons among data points that were in the same chunk than in different chunks, and this was true for both line graphs (where a chunk was defined by connecting dots with a line) and bar graphs (where a chunk was defined by the grouping law of proximity). However, the way line graphs were organized played a much larger role in influencing how participants saw trends than did the way bar graphs were organized.

In addition, some researchers have focused their gaze very narrowly on line graphs per se, trying to understand precisely how they convey information. For example, Shah and Carpenter (1995) showed that readers interpret line graphs primarily in terms of the relative height over the variable plotted on the X axis—and if there was more than one level of the variable represented as a line (and hence there are two or more lines), the participants had difficulty seeing the relation between those levels. For example, one graph depicted the effects of two levels of noise and two room temperatures on achievement scores. When noise level was the variable plotted along the X axis, a cross-over interaction was shown; in this case, both lines tilted down, but one began above the other and ended under it (producing an X pattern, tilting down on one side); the college student participants usually only mentioned the effects of noise on achievement, which was reflected by the fact that both lines went down. When the data were regraphed, with room temperature on the X-axis, the participants now typically mentioned only the effects of room temperature. This finding is consistent with the graphs I presented in chapter 1, showing how plotting the same data in different ways presents very different visual impressions.

Taking a very different tack, Gattis and Holyoak (1996) reported a series of studies of how people use graphs when reasoning. They asked the participants to draw inferences about the relations between two continuous linear variables, temperature and altitude. Specifically, the participants were asked to plot additional data or to decide whether a second line represented a faster or slower rate of change. Gattis and Holyoak found that people could use graphs to reason more effectively when the spatial structure of the graph (e.g., which variables were assigned to the X and Y axis) was congruent with the way a problem was stated (see also Feeney & Webber, 2003, for studies that show effects of the compatibility of text and graphs). In particular, when the independent variable was plotted on the Y axis rather than on the X axis (as usually occurs in all graphs but horizontal bar graphs), participants often misread the slopes. In this situation, they mistakenly assumed that a steeper slope indicates a more rapid increase in the measured amounts. In accordance with the Principle of Appropriate Knowledge, make sure that your readers are familiar with the format you use.

The literature on graph reading, in conjunction with our psychological principles, leads me to make the following recommendations.

Use a line graph if the X axis has an interval scale.

The continuous rise and fall of a line is psychologically compatible with the continuous nature of an interval scale, one that specifies the actual amounts along a continuous measurement scale. The yearly pattern of a young man's fancy is more clearly perceived from **do** than from **don't**. Time, temperature, and amount of money are measured using an interval scale. Moreover, this recommendation is based in part on the findings of Schutz (1961a), who reported that people can determine the nature of a trend better if the data are presented in a line graph than in a bar graph. Provided that the format does not mislead the reader, it makes sense to use the one that is easiest to read.

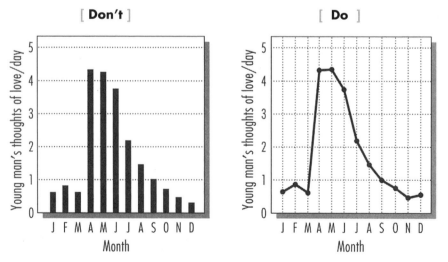

Figure 2.5. The continuous variation of a line is compatible with the continuous variation of time; if you want the reader to note precise point values, put dots or symbols along the line.

Use a line graph to display interactions over two levels on the X axis.

One of the reasons it is so difficult to predict real-world events is that multiple variables interact: The effects of one can be understood only in the context of others. For example, the effects of a presidential election on real estate values in Washington, DC, will depend in part on the type of property. In prosperous sections of the city, values could increase for all but the most expensive homes, whereas in poor sections they may not be affected at all. This interaction, between neighborhood and type of housing, reflects the dependence of the value of one variable on the value of the other.

The units of line graphs—lines and patterns—make them ideal for displaying interactions in the data, because the quantitative pattern is signaled directly by the visual patterns themselves. Most experienced graph readers can immediately recognize that a sideways V pattern (with the ends of two lines converging) indicates one sort of interaction, whereas an X indicates another. As in **do**, we could graph widget production on the Y axis, season (summer vs. winter) on the X axis, and country (Italy, Japan) as separate lines. The crossing lines indicate that in Japan more widgets were produced in summer than in winter, but vice versa in Italy; in contrast, if the lines in this display had formed a V on its side (meeting at the left, over the label "summer," and diverging at the right), that would indicate that there was no difference between the countries in summer, but fewer widgets were made in Italy and more widgets were made in Japan in the winter. In order to see bars as unified patterns that convey specific types of interactions, one must mentally connect the ends of the bars, expending time and effort.

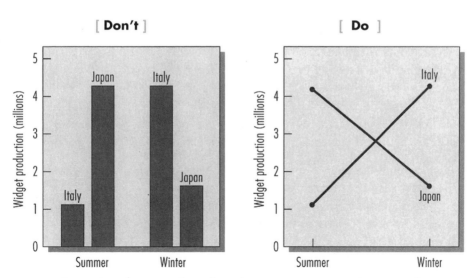

Figure 2.6. Experienced graph readers can interpret typical patterns of lines at a glance.

If there is little chance that readers will improperly infer an interval scale along the X axis, then a line graph is also a good way to convey interactions even if a nominal or ordinal scale is used along the X axis. Nominal scales specify names, not amounts. The numbers on the jerseys of football players are members of a nominal scale. These members often have no necessary ordering (although sometimes an order is imposed by usage), so you are usually free to organize them so that the display is visually simple. Ordinal scales give rank orderings, such as the order in which runners finish a race. Rank-order scales ignore the difference between contiguous values; the difference in time between first and second place may be much larger than between second and third, but this is not indicated in the scale. (For further distinctions among interval, nominal, and ordinal scales, see appendix 1.)

I must mention, however, a study reported by Zacks and Tversky (1999). They showed college students line graphs and bar graphs that depicted height along the Y axis, and the X axis was labeled "Female" and "Male" (gender, a variable with discrete values) or "10-year-olds" and "12-year-olds" (age, a variable with continuous values). Two versions of each graph were created; one with a line connecting the two values, and one with two bars. Zacks and Tversky asked participants to interpret what was depicted in the graph. The participants made more comments that were coded as "discrete comparisons" (e.g., "twelve-year-olds are taller than ten-year-olds") with bar graphs, and more comments that indicate "trend comparisons" with line graphs (e.g., "height increases with age"). (I noticed, however, that there were no tick marks over the labels on the X axis, which can create the visual impression that continuous values are being displayed.) Zacks and Tversky also found that participants tended to draw bar graphs to depict discrete comparisons and tended to draw line graphs to depict trend comparisons.

But one of Zacks and Tversky's (1999) findings gave me pause: A few of their participants (3 of the 25 tested) made trend comparisons with line graphs even when the X-axis contained values on a nominal scale—such as "the more male a person is, the taller he/she is." Even though only a few people apparently read line graphs as always indicating continuous variation, this finding may suggest that line graphs should be avoided even when simple comparisons are made between entries on a nominal scale. Nevertheless, I maintain that line graphs convey interactions more effectively than do bar graphs, and that the risk of such misinterpretations is small compared to the benefits of communicating familiar patterns of relations (e.g., a crossover interaction, signaled by an X-shaped pattern of crossing lines).

Use a line graph when convention defines meaningful patterns.

Patterns of lines can signal specific information for readers who have had experience with similar line graphs (and so have appropriate knowledge). For example, results from the Minnesota Multiphasic Personality Inventory, a standardized personality test, are graphed as scores on subtests. Although these subtests are members of a nominal scale (they have no inherent order), they appear in a conventional order along the X axis. The higher someone scores on each subtest, the higher the point over that location on the X axis; by connecting the points with lines, the graph assumes a jagged appearance. To an expert, the distinctive patterns of peaks and valleys thus formed are signatures of specific maladies.

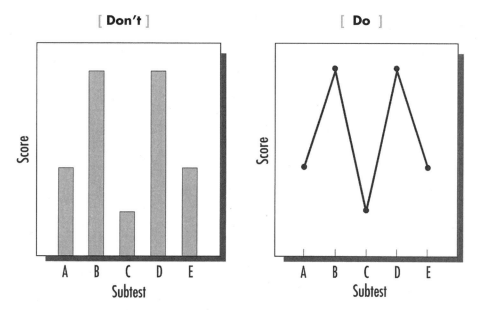

Figure 2.7. If data are presented in a standardized format, readers familiar with the subject can interpret patterns of lines at a glance.

Use a bar graph to show relative point values.

Use a bar graph if the reader is supposed to compare specific measurements. If you want to show how much more one division produced than another for each of three years, you would have three pairs of bars along the X axis—a pair for each year—as shown in **do**. The height of each bar specifies one point value, and this value is relatively easily read. If a line graph is used for the same data, the reader must locate and perceptually isolate a point along a line and note its height along the Y axis; this process violates the Principle of Capacity Limitations—it requires careful allocation of attention and greater effort. You should

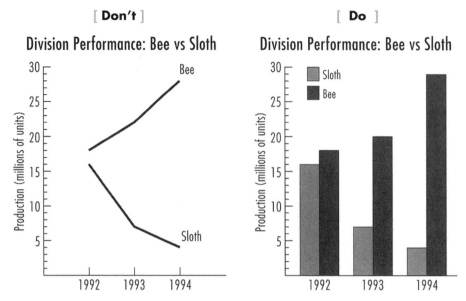

Figure 2.8. The heights of bars define specific points, whereas lines specify continuous variations. It is more difficult perceptually to break up a line into points than to detect the tops of bars.

not require the reader to reorganize a pattern into a new set of perceptual units (in this case, break a line into a set of points) to answer the necessary questions.

Simcox (1983, summarized in Pinker, 1990, pp. 120–122) provided good evidence that bars are encoded in terms of their heights whereas lines are encoded in terms of their slopes. In the most straightforward experiment, participants were asked merely to classify the heights of bars versus the heights of the ends of lines, or were asked to classify the slope or overall height of a line versus the slope or overall height of an implicit line connecting tops of bars. The participants could encode the height at a point faster from bar graphs, and slopes and overall heights faster from lines. For another demonstration that line graphs are harder to use than bar graphs when specific points had to be compared, see Culbertson and Powers (1959). We have found similar results in experiments in my laboratory.

Tufte (1983) suggests that one should never use bar graphs because the bars contain redundant information (too much ink for the amount of data conveyed); he prefers a minimalist graph, illustrating only line segments that correspond to the tops of bars. I reject this opinion for three reasons: (1) The content of the displays Tufte recommends would not be very salient, (2) the heights of line segments are not immediately translated visually into length (a visual dimension we register easily), and (3) Carswell's (1992a) review of the literature on experiments on graph perception did not find support for the utility of Tufte's proposal.

Use a bar graph if more than two values are on an X axis that does not show a continuous scale.

Look at the **don't** version, in which the slope of the line makes it appear as if the income for consulting partners is accelerating rapidly. This visual impression is less striking in **do**. If the intervals of space along the X axis indicate ranks or levels of a nominal scale, not actual equivalent intervals between years, a line will give a misleading impression. In general, if you have three or more levels on an ordinal (ranks) or nominal (names) scale along the X axis, and patterns of lines have not assumed meaning via conventional usage, then a bar graph is appropriate. According to the Principle of Compatibility, the properties of the pattern itself should be consistent with properties of what is symbolized. If a line varies in height continuously, so should the measures of the entities represented along the X axis. The Principles of Compatibility and Informative Changes lead us to use a bar or step graph if the entities do not vary continuously and more than two of them are placed along the X axis.

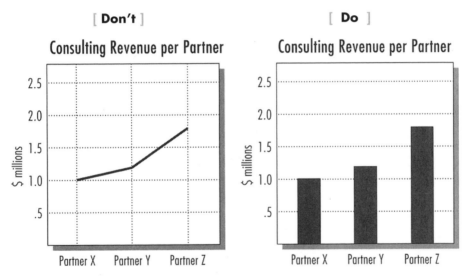

Figure 2.9. A line graph for these data inappropriately suggests a rapid rise along a continuous variation.

Let reality decide between vertical and horizontal bars.

According to the Principle of Compatibility, the properties of the pattern itself should be compatible with what is symbolized. If a graph stands for things that actually have a left-to-right or top-to-bottom order in the world, the graph should symbolize them with bars that have the same spatial arrangement, as in **do**.

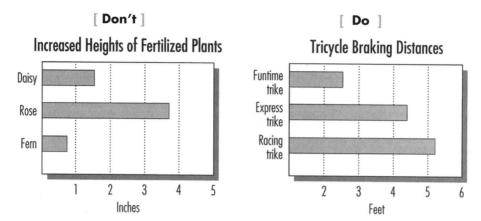

Figure 2.10. The properties of the pattern itself should not conflict with those of the objects or relations being represented.

A mark at the top should always stand for top, and one at the bottom for bottom; one at the left for left, and one at the right for right. Imagine how hard it would be to drive if instead of a steering wheel we had a lever that had to be pushed forward to get the car to turn right and pulled back to get it to turn left.

Use a horizontal bar graph if the labels are too long to fit under a vertical display.

If no other recommendations are violated, use a horizontal bar graph if the labels are long. If a vertical format requires novel abbreviations, the Principle of Relevance can be violated: Readers may receive less information than they need in order to decipher the material effortlessly.

Cleveland (1985) recommends a variant of a horizontal bar graph when each value is labeled. This type of graph has a dot at the location that would be the end of the bar, with a dotted line joining the dot to the Y axis. It is not clear whether such displays are to be preferred over the standard ones.

When in doubt, use a vertical bar-graph format.

Vertical formats are more familiar to most people than are horizontal ones, and increased height hence could be a better indicator of increased amount. Moreover, Schutz (1961a) found that people could determine the nature of trends better when data were presented in a vertical bar-graph format than when they were presented in a horizontal bar-graph format; however, Spence (1990) found no compelling evidence that one format is better the other. In any event, all cultures recognize "higher" as "more," but some cultures, particularly in the Middle East, do not recognize extension from left to right as "more." The common associations of a culture should be respected when designing a display; this is a facet of the

Principle of Compatibility (specifically, the aspect pertaining to cultural conventions).

When to Use a Side-by-Side Graph

A side-by-side graph shows pairs of values that share a central Y axis; its framework is thus a T or an inverted T. Extending to the left are measures of one level of an independent variable (e.g., one type of tax), and extending to the right are measures of another level of that variable (another type of tax). Each pair of bars splayed out from the center illustrates a level of a second independent variable (e.g., country).

Use a side-by-side graph to show contrasting trends between levels of an independent variable.

Look at "Taxes": In which country is the disparity between the two classifications of taxes greatest? A side-by-side bar graph dramatically displays the relative patterns over different levels of an independent variable. As I've noted before, the visual system readily registers differences; readers will quickly note the difference between a regular progression on one side and an irregular one on the other. However, note that the Principle of Appropriate Knowledge may be violated because one set of bars that actually indicates "more" extends greater distances to the left; readers familiar only with Cartesian graphs, in which values to the left of the vertical axis are negative, can be confused. If you have any doubt about your audience's ability to read these displays, don't use them—use adjacent bar graphs instead.

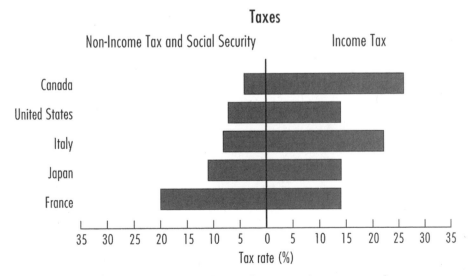

Figure 2.11. Canada has the largest disparity in the two types of taxes—a point that is clear only if a reader knows how to interpret this format.

Use a side-by-side graph if comparisons between individual pairs of values are most important.

Because the bars in a side-by-side graph are immediately juxtaposed, it is easy to notice differences in their lengths as violations of symmetry. The example illustrates the difference in vitamin consumption for two types of guppies; it is immediately obvious in **do** that fan-tailed guppies need disproportionately less B_{12} than other vitamins, compared with spiny-tailed guppies. Side-by-side bar graphs allow the reader not only to contrast trends but also to flag specific differing values. In these displays it is easy to spot the few right/left pairs that are asymmetrical in a sea of symmetries, or the few that have longer bars to the left when the rest have longer bars to the right.

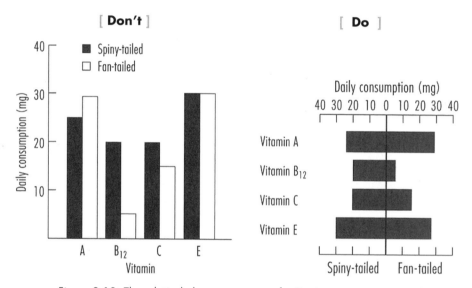

Figure 2.12. The relatively large asymmetry for B_{12} is easy to spot immediately in a side-by-side format.

If differences between paired values are important, illustrate them directly.

Both versions of the "Lateralization" graph illustrate the output from a computer model of the functions of the left and right sides of the brain as the two sides might develop if one has particular experiences and innate biases in how information is processed. In **do** the white portions of the bars indicate the "strength" of certain processes in the left and right sides, and the black portions of the bars indicate the degree to which one side is "stronger" than the other. It is clear at a glance that black bars can extend to the left or right

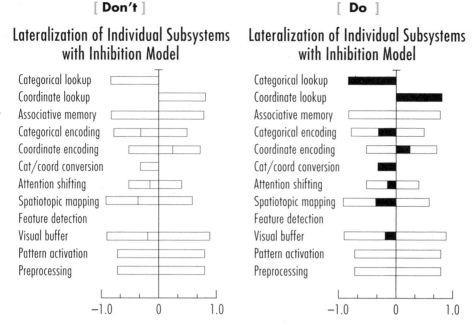

Figure 2.13. The black portions of the bars allow the reader to see immediately the differences in the total extents to the left and right.

side, which indicates that different sides of the brain can be better at specific processes (if the model is accurate), depending on one's personal experiences and processing biases. If your goal is to illustrate a difference in paired values, readers should not be asked to subtract the value of the bars mentally (which would violate the Principle of Capacity Limitations). Why make the readers work harder than they have to (and risk losing them)? It is preferable to present the difference (shown here in black) along with the data.

Avoid using side-by-side displays for more than two independent variables.

Consider a side-by-side display of the average caloric intake (the dependent measure) of male and female gulls for each of twelve months (the independent variables). (See the following page.) Say that months are marked off along the axis, so that there are 12 pairs of bars, and that data from male gulls are represented by bars going to the left and data from female gulls by bars going to the right. This display—the **do** version—is readable. But if we add a third independent variable, age (young vs. old), each bar of the previous display must be replaced by a pair of bars, one for "young" and one for "old" birds. The virtues of side-by-side bars are quickly lost when the arrangement becomes

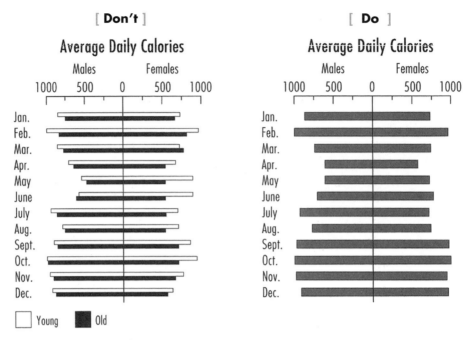

Figure 2.14. If the pattern is too complex, the reader will not easily detect differences in symmetries.

this complicated and we no longer can apprehend at a glance a visual pattern that conveys a message. Unless the data produce regular visual patterns, this format is not well suited for graphing more than two independent variables.

When to Use a Step Graph

Step graphs, like line graphs, use lines as content elements, but these lines trace out horizontal plateaus that change height in discrete jumps (as do the heights of bars in bar graphs). These graphs are like bar graphs in which the bars are pushed together to form a set of steps, and all but the upper line is eliminated. As in bar graphs, the heights of the lines indicate discrete amounts.

Step graphs illustrate quantities, rank orders, or even percentages and proportions (provided that you want to emphasize comparisons among relative amounts, and not the relation of each component to the whole).

Use a step graph to illustrate trends among more than two members of nominal or ordinal scales.

A step graph is useful if the readers are supposed to notice relative changes over three or more values on the X axis but these values do not vary continuously (if they do, use a line graph). In many ways, a step graph is a good surrogate for a line graph for entities that are not arranged on an interval scale: Pushing the bars together creates a pattern that can allow the experienced reader to take in a trend at a glance. The only drawback to these displays is that a bit more effort is required to isolate individual point values, and so a bar graph (where spaces between the bars isolate the point values) is more appropriate if this emphasis is your goal.

Do not use a step graph for two or more variables or levels of a single variable.

From the **don't** graph it is almost impossible to tell the percentage of automobile sales for the different countries. Each independent variable or level of an independent variable is represented by a different series of steps. If the overall values of the two or more levels are similar, the steps will cross and will be difficult to distinguish because of perceptual grouping. Alternatively, if one level has much greater values than another, it will not be im-

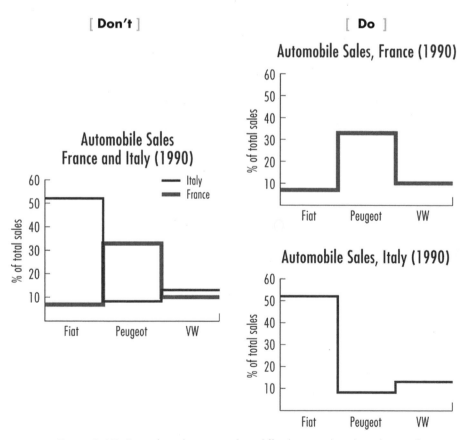

Figure 2.15. A combined step graph is difficult to read and easily mistaken for a layer graph (which would illustrate a cumulative total).

mediately clear whether the upper series of steps corresponds to the cumulative total or simply to the value of one of the levels. In these situations, present the data in separate panels, as shown in **do**.

When to Use a Scatterplot

Scatterplots, which have the standard L-shaped framework, employ point symbols (e.g., dots, small triangles, or squares) as content elements. The height of each point symbol indicates an amount. These displays typically include so many points that they form a cloud; information is conveyed by the shape and the density of the cloud. In fact, Lewandowsky and Spence (1989) report that people read scatterplots more accurately when more points were included, and required comparable amounts of time to read scatterplots with different numbers of points. However, Legge, Gu, and Luebker (1989) found that the ease of reading scatterplots depended only slightly on the number of points. In either event, it is clear that readers get an overall impression of the shape of the cloud. Legge et al. also found that participants could estimate means and variances more easily from scatterplots than from tables of numbers or an alternative display in which luminance was varied to convey amount.

> **Use a scatterplot to convey an overall
> impression of the relation between two variables.**

Scatterplots are appropriate if you want readers to obtain only an overall impression of the relation of two variables. For example, if the heights and weights of 100 people are plotted, you see a cloud drifting up toward the right, indicating that the two variables are positively correlated. You also see that

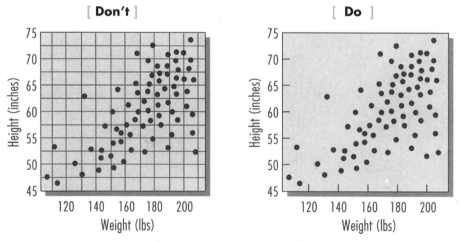

Figure 2.16. Scatterplots convey an impression of trends in the data; they generally are not well suited for conveying values of individual data points, and an inner grid is usually a distraction.

this correlation is not perfect; some very tall people are skinny, and some short ones rotund. Readers who are familiar with these displays can extract a reasonably good sense of how closely values on one variable are related to values on the other. Do not include a grid, as in **don't**; the point is to convey an overall trend, not individual values. Scatterplots often include far more points than can be grouped and compared individually; therefore, they are not a good idea if readers are supposed to discern specific point values (use a table instead).

Use a scatterplot in strict accordance with the Principle of Relevance; it is easy to overwhelm a reader with too much information in these displays. One variant that appears to becoming more common is the box-and-whisker plot. In this type of graph, a box encloses the middle fifty percent of the data and "whiskers" (I-shaped error bars, of the sort discussed shortly) extend to the extremes. A horizontal line in the box indicates the median (see also the "range charts" of Schmid, 1954). In addition, some scatterplots include symbols that vary in size to indicate the number of data points; this variation is intended to allow the reader to see which trends are likely to be most stable (see Bickel, Hammel, & O'Connell, 1975; see also Wainer & Thissen, 1981). However, such augmented scatterplots are often very complex visually, and many readers may not take the time to decode them. Consider your audience carefully before preparing such complex displays.

Avoid illustrating more than one independent variable in a scatterplot.

It is possible to depict different independent variables in a scatterplot by using different symbols to mark the points for different levels of the parameter (see, e.g., Cleveland & McGill, 1984b). For example, say you want

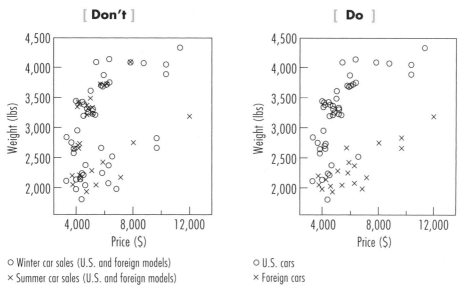

Figure 2.17. Two levels of a parameter can be easily detected in a scatterplot only if their symbols occupy different parts of the display.

to illustrate the relationship between the price of a car and its weight, for U.S. and foreign vehicles. You can clearly see this relationship for both categories in **do**, because the distribution of symbols falls into separate perceptual groups. But data are not always this cooperative: Where the two levels are winter and summer sales of all cars, U.S. and foreign (**don't**), the distribution is such that the different symbols are hard to distinguish and read (and hence the Principle of Discriminability is violated). It usually is better to plot different independent variables in different scatterplots. The only exceptions to this recommendation occur if you want to show that data from two independent variables are intimately related (and so the fact that the points cannot be distinguished is itself compatible with the message being conveyed), or if the clouds formed by different sets of points can be easily distinguished because they are in different parts of the display.

Fit a line through a scatterplot to show how closely two variables are related.

In many cases, scatterplots are used to illustrate a trend, which can be summarized effectively by a line. Unlike the line in a line graph, this "fitted" line does not connect up sets of points. Rather, it falls through the "center"—the most populated area—of the cloud of points. The most popular way to fit such a line requires finding the location that minimizes the average distance of the points to the line (see appendix 1). When such a best-fitting line is provided, it illustrates visually how tightly the points are related by the two variables being graphed. At left is a scatterplot without a best-fitting line; the scatterplot on the right shows the same data with such a line. It is clear that the best-fitting line helps the reader to discern the trends in the data.

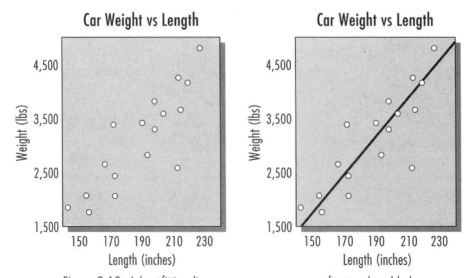

Figure 2.18. A best-fitting line serves as a summary of a trend and helps readers to see the strength of the correlation between two variables.

Graphs for Cumulative Totals

It is sometimes useful to graph cumulative totals, particularly when a quantity varies over time. Stacked bars and layer graphs are used to convey information about how different kinds of things add up to a total, when the total can vary from case to case. For example, one could use segments of a bar to illustrate the amount of taxes paid by the branches of a company in different countries, and have separate bars for each year; the overall height of the bars would change, along with the heights of each segment. Stacked bars are similar to divided bars, the content indicating components of a total amount. Unlike pies and divided bars, which always add up to a constant whole (100%), the cumulative total varies; also, stacked bars often have an explicit L-shaped framework. Like ordinary bar graphs, the framework specifies a measurement scale (on the Y axis) and levels of an independent variable (on the X axis).

Layer graphs (also called layer charts, segmented graphs, or surface graphs) are similar to line graphs, except that the height of each line indicates a cumulative total. They specify how one type of entity and its components vary continuously. The content is a set of layers, each of which indicates the amount of a particular component of the whole. The layers might be the expenditures of different departments in a plant (accounting, production, marketing, etc.). In this case, if the Y axis specified expenditures and the X axis the month, each line would represent the cumulative total for all departments under that line—and each layer (the space between a pair of adjacent lines) would indicate the expenditures from just one department.

When to Use Stacked Bars

Stacked bars consist of two or more parts, each of which specifies the value of a different level of the independent variable. Stacked bars should be used only to illustrate components of wholes that change; they are most effective if the following recommendations are respected.

Use stacked bars if the wholes are levels on a nominal scale.

Stacked-bar graphs have many of the same properties as divided-bar graphs; indeed, the only real differences are that overall height of the stacked bar varies and that typically several stacked bars, illustrating levels of two or more independent variables, are used. The **do** graph uses stacked bars to break down the total number of widgets sold in four cities by industrial and home uses; the overall height indicates total number of widgets. This format is desirable if the independent variable of most

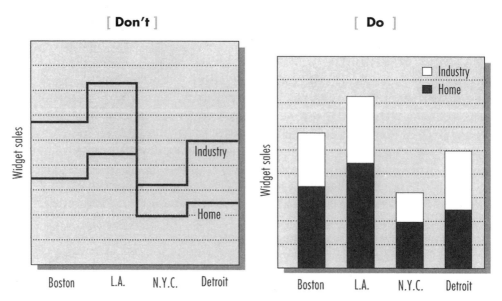

Figure 2.19. Stacked bars clearly illustrate the contributions of each component. The **Don't** arrangement of these data is not easily read and could be mistaken for a step graph.

interest (different cities, in the example) does not specify a continuous quantity.

When to Use a Layer Graph

A layer graph presents cumulative totals. For example, this format could be used to show government assistance to depressed areas in the United Kingdom from 1980 through 1990, divided by four different funding programs. The height of the top line would indicate total funding. This graph would include four layers, formed by the three lines below the line indicating total funding, and each layer would represent the amount of one of the four types of grants.

These displays are like stacked-bar graphs in that they show cumulative totals; they are like line graphs in that they show trends. (In a sense, the stacked bar is a snapshot; the layer graph, a movie.) The following recommendations will help you decide whether you should use a layer graph to display specific data for specific purposes.

Use a layer graph only if the X axis is an interval scale.

If the X axis depicts a continuously changing variable—that is, values on an interval scale—the flux of the lines that define the layers will be compatible with the content, as in **do**. But if the X axis is an ordinal scale (one that specifies ranks) or is a nominal scale (one that names different entities), as in **don't**, the eye will incorrectly interpret the quantitative differences in the slopes of the layers as having meaning (use stacked bars instead). The Principles of Compatibility and Informative Changes suggest that layer graphs be used only if the values on the X axis are arranged along an interval (continuous) scale.

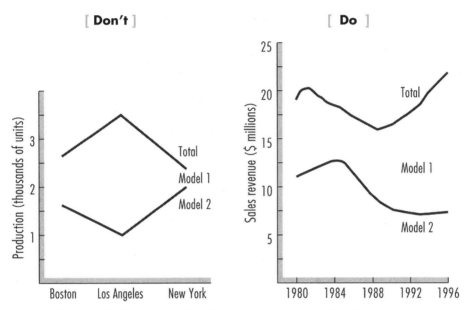

Figure 2.20. The slopes of the lines over a nominal scale—Boston, Los Angeles, New York—misleadingly suggest a continual varying quantity.

Use a layer graph to illustrate changes of parts.

Look at "Trends in Federal Funding": Which was changing more rapidly, the money allocated to basic research or to defense applied research and development (R&D)? Layer graphs are useful if you want to illustrate the change in one component relative to any number of other components with time (or changes in some other interval variable). Because the spaces between lines can be filled, they can be seen as shapes (the visual system organizes regions of common color or intensity into a single shape, as discussed

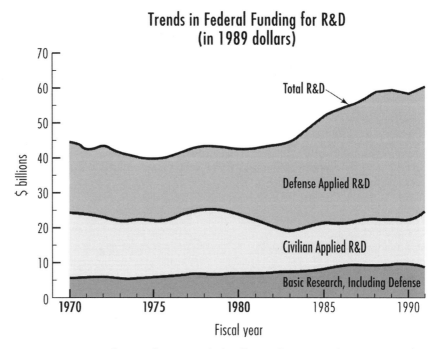

Trends in Federal Funding for R&D (in 1989 dollars)

Figure 2.21. This graph is particularly effective because only one component—Defense Applied R&D—changed dramatically over time, and therefore it is easy to see the differences in the changing widths of the segments.

in more detail in chapter 7). One can easily see a single proportion changing dramatically in a layer graph.

Do not use a layer graph to show precise values of parts.

As with divided-bar graphs, the overall height of the individual lines in a layer graph indicates a cumulative total. If you want readers to extract a specific amount (e.g., money spent on civilian applied R&D during one year, as illustrated in "Federal Funding"), they must note the relative heights of the lines defining a segment at a specific point along the X axis. Given the Principle of Capacity Limitations, this information is difficult to derive; to make a precise comparison, we must subtract heights mentally, at the same time keeping in mind the results of previous subtractions.

When to Use Multiple Panels

Any of the graph formats described in this chapter can be used in multipanel displays, which include two or more graphs. There are two types of multipanel displays. Mixed multipanel displays include two or more different formats; the individual formats are selected independently for each set of data, according to the recommendations offered above. Pure multipanel displays include two or more graphs of the same type. The following recommendations pertain to pure multipanel displays; the individual displays in mixed multipanel displays should be selected in accordance with the previous recommendations. (In chapter 7 we consider how to create both types of displays.) The question becomes, When should you divide the data up and display it in separate panels?

Do not present more than four perceptual units in one panel.

In accordance with the Principle of Capacity Limitations, no more than four bars or clusters of bars, or lines or groups of lines, should be presented in a single panel. If there are more than four bars in a cluster over each point on the X axis, or more than four groups of lines, it is best to divide the

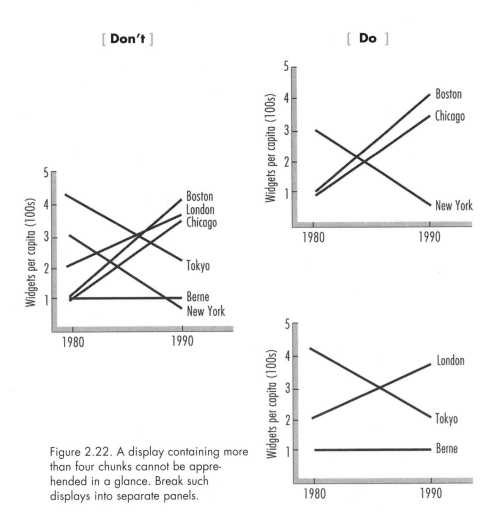

Figure 2.22. A display containing more than four chunks cannot be apprehended in a glance. Break such displays into separate panels.

data into subsets and graph each subset in a separate panel (in chapter 7 we consider how to decide what goes in each panel). If a display is too complicated, we cannot readily understand it because we cannot hold enough of it in mind at the same time (specifically, because of our short-term memory capacity limits), as is the case in **don't**. If a company has six different departments, it is usually better to present separate graphs for, say, U.S. and foreign divisions than to try to cram all that information into a single display. The one exception would be if many of the lines have similar slopes, so that they are perpetually grouped into relatively few units (see pages 12–13).

When we "hold something in mind" we are in fact retaining information in short-term memory. To understand a display fully, one often must draw deductions and notice implications, which requires reasoning; these reasoning processes operate on information in short-term memory. Short-term memory has a relatively small capacity, placing severe limits on what we can understand. Many poor displays are incomprehensible in part because they overload the viewer's short-term memory capacity.

A perceptual unit, or "chunk," is defined as such in accordance with the Principle of Perceptual Organization. Much research indicates that we can hold in mind only about four chunks at the same time; this is an aspect of the Principle of Capacity Limitations (pertaining to our limited short-term memory capacity). Because a chunk is a psychological unit, not a physical one, we cannot determine whether a bar or a line graph should be broken into multiple panels simply by counting marks on the page.

Presenting complex data in multiple panels not only decreases the load on short-term memory but also simplifies the depicted patterns. If the display is too complex, the resulting pattern probably will not be familiar— and so will not convey meaning in a glance. If a complex display is broken into a number of simpler ones, many of the resultant patterns typically can be read without having to note the precise values of individual points. In the following chapters we will consider how to break data into groups, each of which should be plotted in a separate display.

Use multiple panels to highlight specific comparisons.

You may sometimes want readers to compare particular subsets of data. Presenting these data in a separate panel will cause them to be grouped via proximity and hence will lead the reader to compare them. For example, if sales growth is presented for various countries, data for the countries that are to be compared should be plotted together, as in **do**.

This recommendation to use multiple panels to highlight specific comparisons is based in part on results reported by Schutz (1961b), who found that participants could compare lines (determine the one with the highest value at a particular point on the X axis) more easily when the lines were presented in a single display than when they were presented in separate displays. However, there was no difference when participants extracted

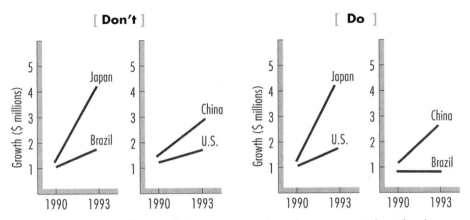

Figure 2.23. The intended comparisons—here, between two industrialized and two nonindustrialized nations—are facilitated by appropriate grouping.

point values from only a single line. The advantage of comparing lines in a single display could have been due to the difficulty of remembering the lines in separate displays. At some point, probably when more than four perceptual units are present, this factor will be overwhelmed by the sheer confusion of too many lines in the display (see also Jarvenpaa & Dickson, 1988).

Optional Features

Finally, consider a number of features that can be included in most of the display formats discussed: a key, error bars, inner grids, the use of two dependent measures, and captions.

Use a key when direct labels cannot be used.

In many cases it will not be clear whether a key is necessary until the rest of the display is produced; only then can you determine whether there is space to label each wedge, object, or segment directly or whether a key would reduce clutter.

To understand a key, the reader must memorize the associations between the names and the content elements, which requires work and may tax our limited-capacity short-term memories. In fact, when Milroy and Poulton (1978) compared direct labels on lines to keys that were placed in the lower right of the display or under the display, they found that readers could use direct labels quickest, without loss of accuracy; they attribute this advantage to fewer processing steps and a lighter load on short-term memory. If at all possible, don't use a key. There are two exceptions to this general recommendation, both of which are illustrated in do: Use a key when

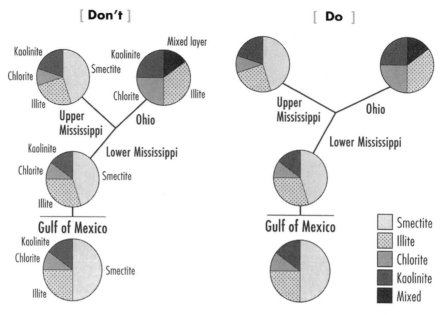

Figure 2.24. A key not only can reduce clutter but also can make a display more visually interesting.

there are so many wedges, objects, or segments in a small space that is impossible to label them directly (the labels would not group properly with what is being labeled or would have to be so small as to be unreadable) or when the same entities appear in more than two graphs of a multipanel display (and hence a key will reduce clutter and the load on short-term memory).

Consider including error bars.

Graphs are often used to illustrate averaged data. An average, by definition, is a measure of the "central tendency" of the numbers. The same average can be obtained from an infinite number of sets of data: 10 is the average of 9 and 11; of 3, 5, 15, and 17; and of −62, −56, 1, 45, 50, and 82. In this case, the variability in the numbers increases for each of the three successive sets (see appendix 1 for a discussion of averages and common ways to compute variability). It often is useful to convey information about the variability in a data set, because readers should have less confidence in the average if the variability is high. In an opinion poll, we have less confidence in the numbers that have large "margins of error" (a measure of variability).

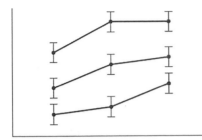
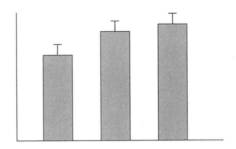

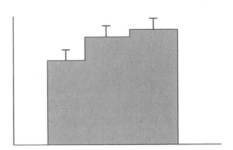

Figure 2.25. Error bars indicate the range over which measurements are likely to vary with repeated sampling, helping the reader to know which differences to take seriously.

If a display is intended to report relative point values or a trend, it is usually a good idea to include, as in the examples shown, a small "I" bar (an error bar) centered on each point, the top and bottom of the bar indicating the range of plus-and-minus one standard error of the mean (or some other measure of variability; see appendix 1).

However, in accordance with the Principle of Relevance, error bars may not be appropriate, depending on your purpose. Error bars were not included in the previous displays, nor will they often appear in later chapters, because of the Principle of Relevance: Because these graphs are designed to illustrate specific recommendations, not to present data, including error bars would provide more information than is appropriate, and their presence would only detract from the point being made.

Use an inner grid when precise values are important.

Use an inner grid if you want the reader to extract specific values, which can be read off the Y axis more easily if one can trace along grid lines. The grid lines serve to connect points along the line or heights of bars to specific places on the axes, taking advantage of the grouping law of proximity to mitigate the imprecision of our visual systems.

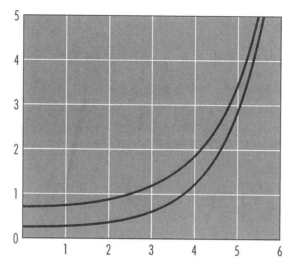

Figure 2.26. Without an inner grid, it is almost impossible to tell that the lines have the same difference in value on the Y axis over the 2 and 5 values on the X axis.

This line graph illustrates a special case in which grid lines are particularly helpful. Look at the vertical difference between the lines over 5 and those over 2; which difference looks larger? In fact, they are the same distance apart, as you can see if you use the grid lines to compare the two differences. Our visual system tends to see the minimal distance between the lines, not their difference in height. Grid lines help the eye to focus on the vertical extent itself (this effect was pointed out by Cleveland, 1985; also see Cleveland & McGill, 1985).

Graph two dependent measures in a single display only if they are highly related and must be compared.

As a rule, do not graph different dependent variables in the same display; in most cases, graph the different data in separate panels of a multipanel display. Graphing two dependent variables in the same display forces the reader to keep track of what goes with what, taxing our limited processing capacity. The only exception to this recommendation occurs when the dependent variables are intimately related and their interrelations are critical to the message being conveyed, because plotting the data in the same display allows the laws of perceptual grouping to organize them into a single pattern. The volume of oil pumped and the failure rate of a piece of pumping equipment are probably connected, and it makes sense to graph these two dependent measures in one display, as in **do**. On the other hand, oil volume and the CEO's frequent-flier mileage (**don't**) are not as immediately related, and the graph is not particularly helpful.

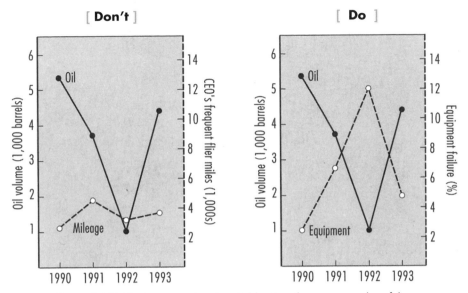

Figure 2.27. Plotting two related variables together can reveal useful information, such as the impact of equipment failure on the amount of oil that was pumped.

Do not use mixed bar/line displays to show interactions.

As is evident in **don't**, it is more difficult to see interactions if a mixed display is used because bars and lines do not group to form simple patterns

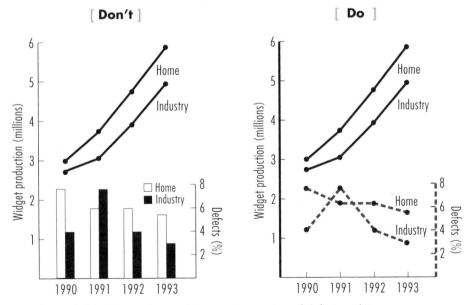

Figure 2.28. The trend toward decreasing number of defects with increasing production, especially for the industrial widget, is not immediately evident from the mixed line/bar display; the visual system more easily groups different lines into a pattern than it groups lines and bars.

(they are too dissimilar, and so the Principle of Perceptual Organization [grouping rule of similarity] will not be effective); for easiest comparison of trends and interactions, use multiple lines.

Include a caption to clarify unfamiliar terms and specific features.

A caption is an explanatory note accompanying the display. The Principle of Appropriate Knowledge leads you to ensure that the reader knows the terms necessary to understand the graph, and in some cases labels should be clarified in a caption. If the display is complex, note explicitly in the caption which patterns in the display are of most interest.

The Next Step

The next step is to begin to create the display. If you have selected a pie graph, divided-bar graph, or visual table, go to chapter 4; if you have selected any of the other types, which have an L-shaped or T-shaped framework, go to chapter 3.

Creating the Framework, Labels, and Title

3

THE DIFFERENT GRAPH formats share many features. All have labels and a title; most (except pie graphs, divided-bar graphs, and visual tables, discussed in chapter 4) have an L- or T-shaped framework. In this chapter, recommendations are presented for creating the framework, labeling it, and titling it. Once the framework is labeled and titled, turn to the appropriate chapter—4, 5, or 6—to create the content elements; to label them, follow the recommendations in this chapter.

Creating the L-Shaped or T-Shaped Framework

The framework of a bar, line, stacked-bar, step, layer, or scatterplot graph is an L-shape. The Y (vertical) axis indicates the amount of what was measured (the dependent variable), and the X (horizontal) axis indicates the things to which the measurements apply (levels on an independent variable); the roles of the vertical and horizontal legs are reversed in horizontal bar graphs. Side-by-side graphs, which are like two horizontal graphs that share the same Y axis, have a T-shaped framework. First we consider general recommendations about the overall form of the framework; we then turn to the details of producing the X and Y axes.

Ensure that the framework is easily detectable.

According to the Principle of Discriminability, marks must be large enough and drawn with lines that are easily seen, as in **do**. Neurons that detect changes in one region of space inhibit other neurons that detect changes in nearby regions, and so if a mark is not large or heavy enough, the cells that typically would detect it are inhibited from responding. Detectability depends on a wide range of factors and their interactions, such as the color of the figure and its background, the contrast of intensity of light and dark elements, and the shape of the figure. The very large number of possible interactions makes it impossible to provide hard-and-fast rules to ensure that a mark can be noticed.

[**Don't**]   [**Do**]

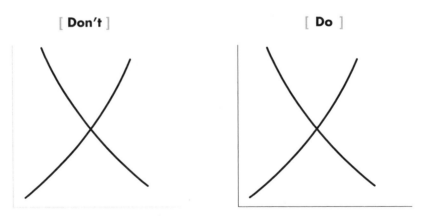

Figure 3.1. If the framework is difficult to discern, the content functions as a visual table.

Group the parts of the framework to form a single unit.

The legs, or axes, should be connected; then by the laws of perceptual grouping the framework will be seen as a single whole, as in **do**. If the legs are not connected, and the gap is too large to be completed perceptually by good continuation, the reader will have to work to figure out the display.

[**Don't**]   [**Do**]

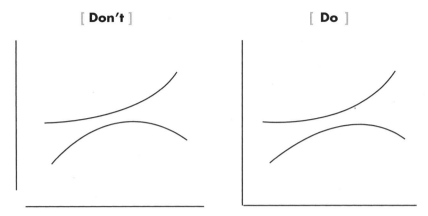

Figure 3.2. Confusion can ensue if the segments of the framework do not form a single unit.

Ensure that the height-to-width ratio of the axes makes differences in content discriminable.

The ratio of the height to the width is the *aspect ratio.* If the X axis is too long relative to the Y axis, differences in bar heights (or line heights, for line graphs) may not be readily discriminable. All bars drawn in the **don't** framework would be relatively small, and the percentage of difference not large enough to be easily discriminated. Adjust the aspect ratio so that, in accordance with the Principle of Compatibility, actual differences in the data produce corresponding visible differences in the display.

[**Don't**]   [**Do**]

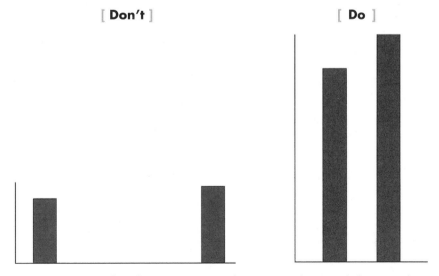

Figure 3.3. Adjust the aspect ratio so that, in accordance with the Principle of Compatibility, significant differences in the data produce corresponding visible differences in the display.

For accurate visual impressions in a line graph, duplicate the Y axis at the right.

A curious visual illusion causes people to underestimate the right-hand point along a line if it has to be compared against a Y axis on the left-hand side; this misapprehension is amplified if two or more lines are graphed in the same framework (see Hotopf, 1966; Poulton, 1985; see also Hotopf & Hibberd, 1989). The error is not very large—at most, about eight percent—and may not be important for many purposes. But if you want the reader to gain an accurate visual impression, draw the Y axis on both sides of the graph, as in **do**. For aesthetic purposes, you can always "box" an L-shaped framework, as in **do**.

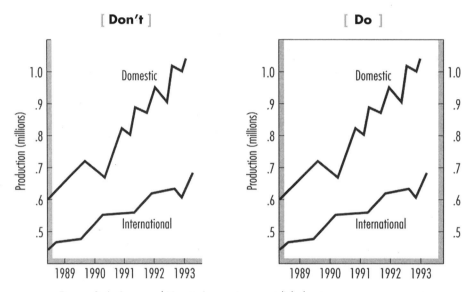

Figure 3.4. A second Y axis (or an inner grid) helps to convey an accurate visual impression of the relation of two lines at the right side of a graph.

In addition, it should be noted that our visual systems tend to distort the slopes of rising content lines so that they are closer to a forty-five degree diagonal; we overestimate slopes of relatively shallow lines and underestimate slopes of relatively steep lines (Schiano & Tversky, 1992; Tversky & Schiano, 1989). This phenomenon is good reason to include an inner grid (discussed in chapter 2) if you want readers to have an accurate impression of changes. Note that the use of an inner grid in general is a good idea if you want the reader to obtain specific point values; moreover, if you use an inner grid, there is no need for a second Y axis.

If two dependent measures are plotted, use different colors or patterns for the two Y axes.

As noted in chapter 2, you should only present two different dependent measures, each on a separate Y axis, in the same display when you want to emphasize a close relation between them. Line graphs are usually the most appropriate type of graph for such information because lines produce the most easily identifiable patterns. In accordance with the grouping law of similarity, use the same color or pattern to plot the content line and the corresponding Y axis, as in **do**, allowing the viewer to see immediately which Y axis goes with which data. When using colors, choose ones that are well separated in the spectrum, and make sure that adjacent colors have different levels of brightness. Use warm colors to define a foreground, avoid using red and blue in adjacent regions, and avoid using blue if the display is to be photocopied. (For a further discussion of color, see chapter 7.)

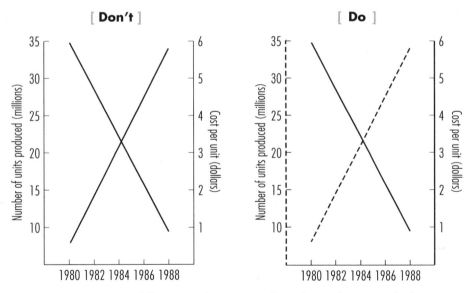

Figure 3.5. A difference in the patterns allows the reader to see which Y axis goes with which content line.

**With two dependent variables, put one Y axis
on the left and one on the right.**

To promote discrimination and proper grouping with the content, put
one Y axis on the left and one on the right, as in **do**, rather than putting
both on the same side; putting both axes in the same area can result in incorrect grouping of labels with axes.

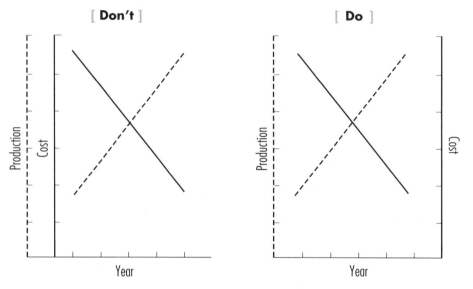

Figure 3.6. Proximity improperly groups two Y axes on the same side of the
graph with each other and the inner axis with both content lines.

Creating the Axes

You need to address four questions in order to produce the appropriate X
and Y axes:

1. If there is more than one independent variable (e.g., age and income
   level), which one should label the X axis, and which should be
   treated as parameters, represented by separate bars or lines?
2. How should the levels of the variables along the axes be ordered?
3. What range of values should be included along the framework?
4. How should tick marks be produced?

For convenience, in the following recommendations I will assume that
the X axis is used to present an independent variable (company, country,
political party, etc.) and the Y axis is used to present the dependent variable

(dollars, amount of oil, number of votes, etc.). In horizontal graphs, these roles are reversed; if your format is horizontal, treat recommendations for the X axis as recommendations for the Y axis, and vice versa.

**Put the most important independent variable
on the X-axis, and treat the others as parameters.**

Which team had more scored points, the Giants or Redskins? This should be easier to determine from the graph on the left. But if you want to know how the points were made—which team had more scored points than allowed points—the version shown on the right is preferable. The difference between the two is the choice of independent variable to put on the X axis and the corresponding parameters (the different bars). Because our visual systems group things that are close to each other, the two graphs make different sorts of information easiest to detect at a glance. The choice of which independent variable to put on the X axis depends in large part on the purpose of the display: The most important independent variable should be on the X axis. That variable will be perceived as the most important factor, and the others will be perceived as secondary.

To determine which variable is the most important, write down a concise description of what you take to be the main point of your graph. In the football example, perhaps your summary is: "The Giants had roughly equal numbers of scored and allowed points, whereas the Redskins had many more scored points." Here the important contrast is between teams, which is the subject of the sentence and the first element mentioned, and your graph would look like the one on the left. On the other hand, you might write:

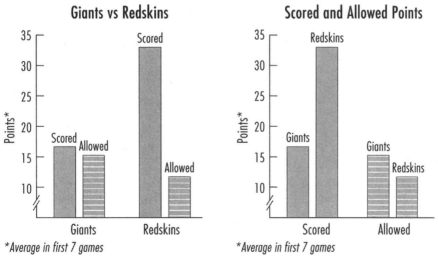

Figure 3.7. Depending on the variable chosen for the X axis, different patterns emerge and different trends and interactions are easy to see.

"More scored points were made by the Redskins than by the Giants, whereas more allowed points were made by the Giants than the Redskins." Here the focus is on the difference between scored and allowed points, suggesting that this is the most important variable, which would lead you to use the arrangement of the graph on the right.

This recommendation is based on the Principles of Relevance and Capacity Limitations. If an inappropriate organization is used, the reader must mentally transform the graph to extract the information you are attempting to highlight; the reader can easily see differences between bars at adjacent points along the X axis, but must work to locate distant corresponding bars and remember what to compare with what. This effect is even more dramatic with line graphs, where the orientation of the line conveys information about variations of levels of the independent variable on the X axis, and differences in the relative heights of the lines at each point convey information about the variations of levels of the parameters (see pages 18–20).

When in doubt, put an independent variable with an interval scale on the X axis.

If the independent variables are equally important, is one of them on an interval (i.e., a continuously varying) scale? If so, put that variable on the X axis, as in **do**. The progressive variation in heights from left to right

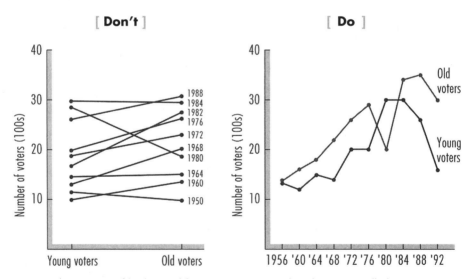

Figure 3.8. If both variables are on an interval scale, it is usually best to put the one with more levels on the X axis. This choice produces fewer lines and, usually, a simpler pattern.

along the X axis will then be compatible with the variation in the scale itself; such changes in height are easily detected by our visual systems (unlike detection of differences of differences in height, which we do not register easily, as shown in chapter 1). If there is more than one independent variable with an interval scale, it usually is best to put the one with the greatest number of levels along the X axis. For example, if you are graphing the number of people of different ages who voted Republican in Somewhere County in different years, put year on the X axis if you have ten years and only two age groups, and vice versa if you have only two years and ten age groups. This procedure cuts down the number of separate content elements and thereby reduces the load on our short-term memory capacities.

All else being equal, put on the X axis the variable that produces the simplest pattern.

If none of the foregoing recommendations applies to a particular case, put on the X axis the independent variable that allows you to make the simplest pattern of content elements; this arrangement will be the easiest to apprehend, as you can see by comparing **don't** and **do**. This recommendation is easiest to carry out if you use a computer to create the graph and can experiment with different ways of producing the display.

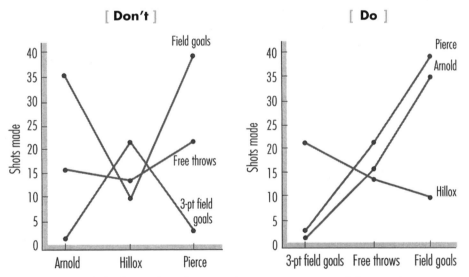

Figure 3.9. The number of crossing lines is one measure of the complexity of a display. However, if crossing lines form a familiar pattern, they can be comprehended more easily than unfamiliar uncrossed lines.

Creating the Scales

Once you have decided which independent variable to place along the X axis, you must order the levels of that variable. If the levels reflect measurements along an interval or ordinal scale (years, weight, age, placement in a race), the order is obvious: the levels are ordered from smallest to largest (left to right or bottom to top). However, if the levels are names of things (e.g., countries, players, products, etc.) that are not drawn from an interval or ordinal scale, your decision about how to order them is more complex. Perhaps the levels lend themselves to a conventional or commonsense order, as in the lifetime income of members of the rock group Crosby, Stills, and Nash. If they don't, consider the following recommendations.

Use position to indicate greater or lesser quantities.

Spatial dimensions have a psychological beginning, symbolized in a graph framework by the physical intersection of the X and Y axes, by convention the site of the origin. The greater the distance from that beginning point, the larger the indicated amount. The Principle of Compatibility suggests that quantities (e.g., time or volume) should increase from bottom to top, left to right along a line (as in **do**).

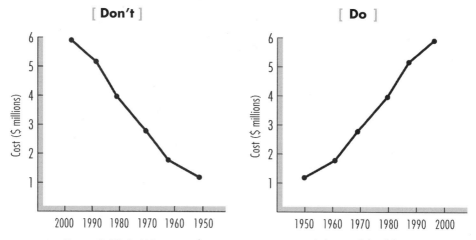

Figure 3.10. In Western culture, a pattern moving left to right will be automatically interpreted to indicate "more."

Make display locations consistent with locations they represent.

The Principle of Compatibility suggests that data about actual positions in space should be presented in the corresponding positions in the display, as in **do**. I have (too often) seen this principle violated in graphs showing properties of the left and right sides of the brain, where "left" is on the right side of the X axis and "right" is on the left side. This forces the reader to reorganize the display mentally, a task that requires effort and is sometimes prone to error.

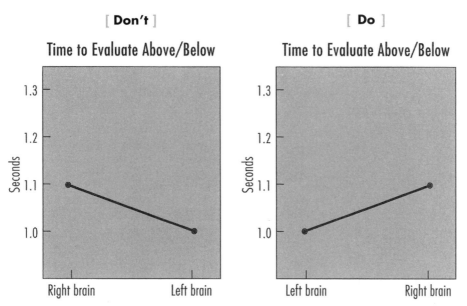

Figure 3.11. The left side of something should be illustrated at the left end of the X axis, and the right side should be illustrated at the right end of the X axis.

Organize a nominal scale so that cases to be compared are adjacent, or so that the simplest pattern is produced.

Both **don't** and **do** illustrate the percentage of 18- to 19-year-olds who were full-time students during the academic year 1987–88. The **do** version is much easier to take in at a glance: You immediately see the ordering and differences among the countries. If you want the reader to compare specific cases—that is, levels on a nominal scale—put their steps or bars next to each other (Spence & Lewandowsky, 1991, found that adjacency helped people compare bars but did not matter very much when they compared wedges of a pie graph). If no specific comparisons are particularly important, the Principle of Informative Changes leads me to recommend that a simple progression of increasing height from left to right is to be preferred over a jagged series, if the ordering along the X axis is arbitrary. If you don't do this, the reader may assume that each jag means more than it actually does and waste time figuring out what it's supposed to mean (here the only meaning is that some things have more than others, differences measured by the heights of the steps).

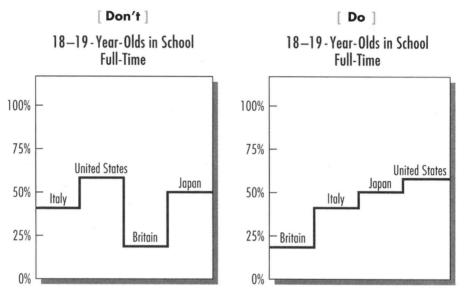

Figure 3.12. If no particular comparison is especially important, order the elements in a simple progression.

Adjust the range of values on the Y axis so that the visual impression reflects the appropriate information.

What range of values should you demarcate along the Y axis? You may have data that range from 100 to 300 but the possible range is 0 to 500. In most cases, it is up to you to decide what range of values to include.

In which version, **don't** or **do**, are you more likely to notice that more blue-collar families were in the market for big-ticket items in 1991 than in 1990, but vice versa for white-collar families? Visually, the difference between the pairs of bars is more striking in **do** because the scale is not a uniform progression from zero. If it were, as it is in **don't**, the significant differences would be less apparent.

According to the Principle of Compatibility, the visual impression produced by a display should convey actual differences and patterns in the data. To produce an accurate impression, display only the relevant range of the scale along the Y axis. Unless the zero value is inherently important, make the visible scale begin at a value slightly lower than the smallest value in the

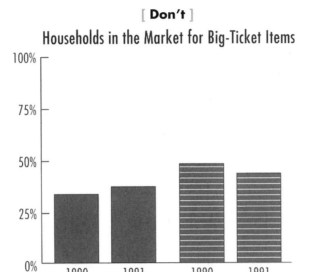

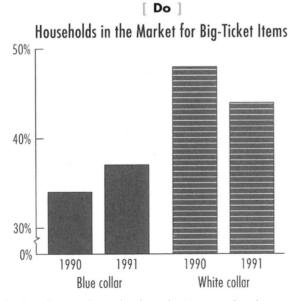

Figure 3.13. Adjusting the scale along the Y axis makes the opposite trends for blue-collar and white-collar households immediately evident. A discontinuity in the scale can be indicated by slash marks or a zigzag on the axis.

data, and the upper value slightly larger than the largest value. To show that you are leaving out a portion of the scale, make a clear gap in the Y axis (with two short slash lines or a zigzag, which will draw the reader's attention to this deletion) slightly above the origin. The gap in the axis clearly signals a discontinuity in the scale along the Y axis. Adjusting the scale in this way allows readers to discriminate the important differences in the height of the content elements, which reflect the important differences in the data.

Cleveland (1985) worries that bars are misleading if the initial value is not 0 (and hence instead suggests using dotted lines that lead to a filled circle marking the total amount); in my view, this is not a problem: The important consideration is that the visual impression of differences and trends is compatible with the actual differences in trends in the data. Provided that this is true, and marks are used to indicate discontinuities in a scale, there is no reason not to use bars when part of the Y axis has been excised.

You may also want to consider excising the middle of the scale if you want the reader to see a difference in trends that is significant in the data, if you do not care about conveying the magnitude of the average differences between levels, or if the difference in trends would be difficult to see if the entire scale were presented because there is a large region of the scale in which there are no values plotted. Such excision will eliminate the unused portion of the scale, in essence producing two graphs, one above the other. This practice is unusual, however, so it is especially important that the slashes used to mark the break in the Y axis be easily detected.

Transform the scale to convey the appropriate visual impression.

Both of the graphs to the right present the same data, but they tell different stories. The top one obscures the differences among the methods. In contrast, the bottom graph reveals the key differences in the resolutions of the different methods. If in fact there were no real—that is, statistically significant—differences in the sensitivities of the methods, the bottom display would be an example of graphical deceit. When there really are significant differences, however, the Principle of Compatibility leads us to consider transforming the scale so that the visual impression reflects the actual pattern in the data. Is the range of measurements so great that important variations will be lost if a uniform scale is used? Do differences tend to increase with larger values? Is your purpose to illustrate trends or relative differences among similar cases, not the actual differences in overall amounts? If so, consider using a logarithmic scale on the X or Y axis, or both, as appropriate.

A logarithm is an exponent. When, as often, it is based on 10, it indicates the power to which 10 would have to be raised to produce a certain number. The log of 10 is 1, the log of 100 is 2, and so forth. The same actual distance on a logarithmic scale represents increasing amounts as one moves farther along the scale. Each equal increment represents multiplying the value of the previous increment by another factor of 10: If the first ticked-off space along the Y axis marked off values from 0 to 10, the next would specify values 10 to 100, and the next values 100 to 1,000. Thus, logarithmic scales compress differences among large numbers, because the same physical distance on the scale stands for increasingly large increments as the num-

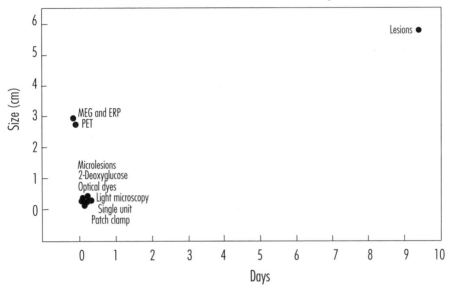

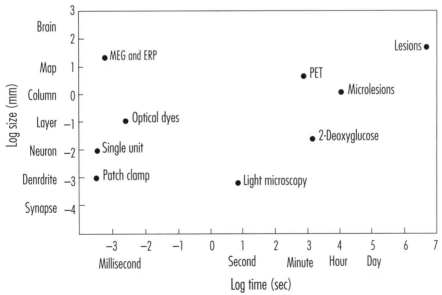

Figure 3.14. Using logarithmic scales here allows the reader to see differences among the higher resolution techniques that otherwise are completely obscured.

bers get larger. Using a logarithmic scale is desirable if doing so makes important variations more discriminable, or if there is a great range of values on the axis and differences increase with larger values. If the scale is transformed, it is critical that it be labeled appropriately. (See Cleveland, 1985, for a more detailed discussion of logarithms and their use in graphs; Cleveland & McGill, 1985, recommend using base 2 logarithms if fractions of exponents would be necessary with base 10; they also point out that base 2 makes increasingly good sense as more people use computers and develop intuitions about powers of 2, but the norm currently is still base 10.)

Do not abbreviate or transform the scale on the Y axis in a stacked-bar graph.

The main point of a stacked-bar graph is to convey information about the relative proportions of different components of the whole. Excising part of the scale on the Y axis, as in **don't**, or using logarithms or another nonlinear transformation, will alter the visual impression so that the sizes of the segments no longer reflect the relative proportions of the components; the visual appearance will not be compatible with what is symbolized.

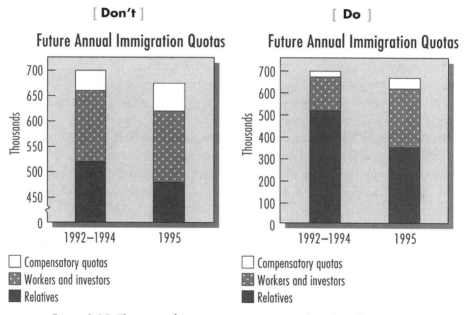

Figure 3.15. The sizes of segments are meant to reflect the relative proportions of components; if the scale is adjusted, the visual impression is necessarily misleading.

Arrange interval values on the X axis of a line graph at distances proportional to their magnitude.

The sharp increase in consulting revenues at Arthur Andersen, Inc., implied in **don't**, did not in fact occur. Variations in the heights of lines suggest variations over continuous quantities when three or more points are placed along the X axis. The Principle of Compatibility implies that we should arrange the values at distances proportional to their actual values. If you are graphing number of ulcers for employees who have been on the job for one, two, and four years, the tick mark for year 4 should be placed at a distance from year 2 that is twice that between year 1 and year 2. Otherwise, the rate of rise and fall will not accurately reflect the trends in the data.

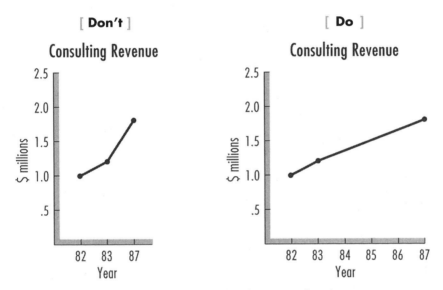

Figure 3.16. If values on an interval scale are not placed at correct proportional distances along the X axis, the content line will misrepresent trends in the data.

Graph positive and negative values relative to the zero baseline.

From which graph is it more obvious that Japan had the largest negative change in this composite index of leading economic indicators? In which case is it more obvious that the economies of Japan and the United States contracted, whereas those of Germany and Italy grew slightly? A graph like **do** is useful if there is a negative element to be portrayed; in this case there

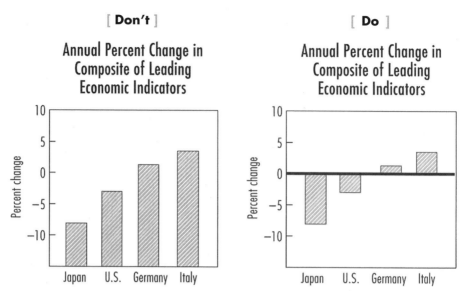

Figure 3.17. When positive and negative values are compared, make sure that the zero baseline is salient.

were years in which contraction, not growth, occurred. On such a graph the zero baseline should be indicated by a heavy line in the middle of the axis so that some bars hang below the zero origin. Our visual systems notice differences, and this type of display makes such differences explicit.

Use reference lines to mark watershed points.

If specific values correspond to watershed points, it may be useful to include a reference line to indicate these points. For example, if you are graphing the average weight of a herd of cows for each day over a period of a year, you might want to insert vertical reference lines every two months, when the type of feed was changed. These lines should be clearly labeled on the X axis (see Cleveland, 1985, for further discussion of the utility of such reference lines).

In a T-shaped framework, mark the X axis using the same scale on both sides of the Y axis.

The two segments of the horizontal line that is the X axis function as two scales for the dependent measure, one extending to the left of the vertical Y axis and one to the right. In keeping with the Principle of Compatibility, the same visual extent of bars on the two sides should signal the same amount, as in **do**. Similarly, to make it easy to compare the bars in each pair,

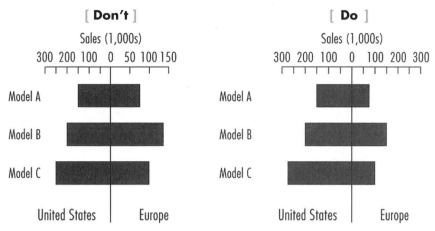

Figure 3.18. The visual impression of the disparities in U.S. and European sales does not reflect the actual disparities in **Don't** because different scales are used for each of the two regions.

give the left and right bars the same origin. If you do not, you may create a misleading visual impression.

Creating the Tick Marks

Tick marks indicate increments along a scale; they demarcate steps of increased amount (along an interval or ordinal scale) or the presence of a particular entity (along a nominal scale).

Extend tick marks inside the axis.

Having the tick extend inside the axis helps to group the label perceptually with the corresponding portion of the content.

Place tick marks at regular intervals.

Remember that visible differences not only attract the eye but also lead the reader to expect additional information; if there is no such information forthcoming, do not signal it. One notable exception to this general recommendation sometimes occurs with a logarithmic scale; in that case, an increasingly dense placement of tick marks signals that the axis is calibrated in this scale.

Place larger ticks at labeled values and halfway between labeled values.

If you want the reader to be able to estimate values that fall between labeled values, put a larger mark at the labeled value and at the value halfway between labeled values. For example, if the X axis labels intervals of ten years, place a heavy tick mark at each ten-year mark and a slightly smaller one at the five-year increments, as in **do**. However, if intermediate point values are not important, make all the intermediate ticks the same size and weight (Principles of Informative Changes and of Relevance are key here).

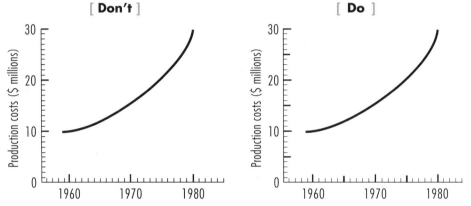

Figure 3.19. Intermediate heavier tick marks help the reader to estimate values along the content. An inner grid would also be appropriate if the reader is meant to extract relatively precise values along the trend.

Do not place ticks or labels on the Y axis of a T-shaped framework.

In a T-shaped framework, the Y axis is the central vertical line; it should not have ticks or any direct labels. In a horizontal bar graph, the labels are placed along the Y axis to promote grouping with the bars; in a graph with a T-shaped framework (a side-by-side graph), a second bar appears in the space that would otherwise be used for a label.

Use tick marks if specific values are important.

In accordance with the Principle of Relevance, do not use tick marks if your aim is to portray a general trend, not to provide actual measurements. Both versions of the Microsoft graph are meant to show that the value of stock increased dramatically relative to the Dow Jones Software Index. For this purpose, the version on the right works better than the one on the left, where unimportant fluctuations serve only to distract the reader.

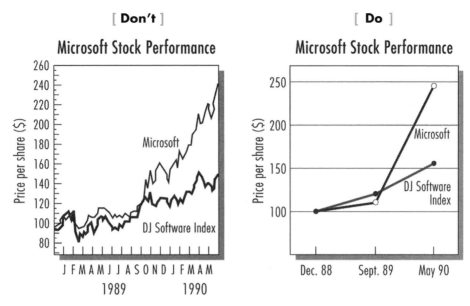

Figure 3.20. To portray a general trend, do not supply numerous tick marks or extraneous detail in the content.

Creating the Labels

If the labels are too small, are grouped improperly, or otherwise violate principles of perception and cognition, the display may be uninterpretable. The following recommendations apply to all labels in a display.

Ensure that labels are easily detectable.

How often have you seen a display like **don't**? The designer apparently forgot that the display was going to be reduced! Ensure that labels and all other parts of the display are clearly discernible—and will remain so even if the size of the display is reduced before reaching the audience. Check whether your display becomes illegible by reducing it in your computer graphics program or using a photocopier with a scaling feature.

The most common violation of the Principle of Discriminability involves size. Optimum size depends on two factors, the absolute size of the characters and the distance of the reader. Thus, the "best" size for legibility is expressed as a relationship: the height of the characters divided by the distance from the reader. In studies, participants were asked to move toward a display until they could just read it; measurements were taken at this point and the ratio calculated. The ratio is expressed in radians, the unit of the proportion

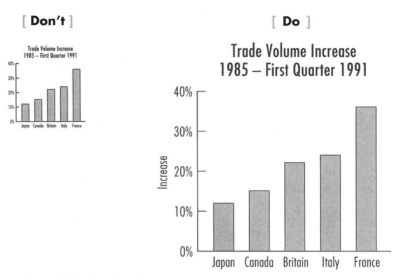

Figure 3.21. Plan ahead!

of an arc of a circle (in effect, the height of the character) to the radius of that circle (the distance of the reader from the display). The U.S. military requires letters at least 0.007 radians, the proportion of size to distance at which people can read virtually perfectly. This size corresponds to 24 minutes of arc; a minute of visual arc is 1/60th of a degree, where a circle is divided into 360 degrees. To get a sense of this size, take a look at one of your thumbs when you hold it up at arm's length—your thumb subtends about 2 degrees.

Specifically, Smith (1979) reports a large-scale study of legibility of letters, conducted under naturalistic conditions; the stimuli ranged from individual letters to words and text and varied in typeface as well as size. Smith reports that ninety-eight percent of letters can, in general, be read if the minimal size is 0.0046 radians, and that people read virtually perfectly when the letters appear at 0.007 radians. However, the precise size at which letters are easily read depends on a host of factors, ranging from the typefont to the level of illumination. Thus, you must be aware of this potential problem and ensure that the labels in your particular display are large enough to be easily read.

Avoid fonts in which letters share many features.

It is easier to distinguish a dog from a cat than a wolf from a German shepherd dog. The more characteristics that two objects have in common, the more information the visual system must register before it notes a difference. The same is true for letters in different fonts. Although the terminology varies widely, I will use the term "font" as the most general category that includes all styles of print. Within this category, fonts can vary in terms of their typeface (e.g., Times or Arial), style (e.g., *italic*, **bold**, UPPERCASE), size, and color. Fonts in which letters share many features (e.g., vertical lines, horizontal lines, specific diagonal lines) are difficult to read because the reader must look carefully at each letter. For example, many researchers have found that IT IS DIFFICULT TO READ TEXT PRESENTED ONLY IN UPPERCASE. Using mixed upper- and lowercase fonts enhances not only reading speed but also comprehension (e.g., Wheildon, 1995). Tinker (1963) reports that people read words in all uppercase ten to fifteen percent more slowly than words in all lowercase. Moreover, *italics are more difficult to read than plain text* (e.g., Hill, 1997; Tinker, 1963).

Use visually simple fonts.

The desire to make an attractive display presents the temptation, succumbed to in **don't**, to use a "fancy" font. Such fonts are difficult to read, and so you should avoid them. But what counts as a "fancy" font? Typefaces are divided into two general classes, *serif* and *sans serif*. Serif typefaces have little hooks, feet, and brackets (*serifs*) at the ends of line segments; sans serif typefaces lack these flourishes and are composed only of straightforward strokes. For example, serif fonts include Times, Times New Roman, Palatino, Garamond, and Century Schoolbook, and sans-serif fonts include Arial, **Helvetica**, Futura, Tahoma, Avant-Garde, Century Gothic, and Verdana. I have sometimes heard or read (e.g., on page 13 of the style guide *The Basic Elements of Design*, which came with an early Apple LaserWriter) that titles and labels should be in typefaces without serifs. Fonts with serifs

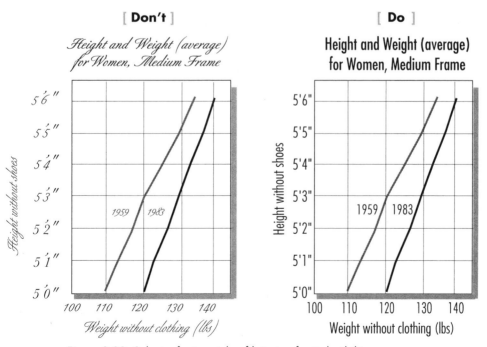

Figure 3.22. Select a font or style of lettering for its legibility.

are said to be more complex than those that do not have serifs, such as Helvetica. Serifs are supposed to help you read words more quickly by providing additional cues, but some believe that they actually make letters less discriminable.

An enormous amount of ink has been spilled, both literally and electronically, on the question of whether serif or sans-serif typefaces are more easily read, and people hold surprisingly strong opinions. For example, we read "Sans-serif fonts are definitely more readable than Serif fonts" (De Rossi, 2002). In contrast, we can also read "Serif type is more *readable* and is best for text . . ." (Williams, 2003, p. 63).

Many studies have been conducted on this topic. For example, Bernard, Mills, Peterson, and Storrer (2001) studied the time people required to read text in 12 different fonts (some serif, some sans serif) on a computer screen. The participants also rated how much they liked the fonts. These researchers concluded that "no significant difference [sic] in actual legibility between the font types were detected. There were, however, significant differences in reading time, but these differences may not be that meaningful for most online text because these differences were not substantial." They did claim, however, that Courier, Comic, Verdana, Georgia, and Times were perceived as being most legible, and participants preferred Arial, Comic, Tahoma, Verdana, Courier, Georgia, and Schoolbook typefaces.

Tullis, Boynton, and Hersch (1995) asked participants to proofread text and investigated the speed of reading different typefaces (on a computer screen). They examined Arial, MS Sans Serif, and MS Serif at 6, 7, 8, 9, and 10-point font sizes. The participants read the serif and sans-serif fonts equally quickly but read the two larger fonts sizes more quickly. Gould, Alfaro, Finn, Haupt, and Minuto (1987), Smedshammer et al. (1990, reviewed in Frenckner, 1990), and Zachrisson (1965) found no differences in reading speed for the two kinds of typefaces. Tullis et al. (1995) also collected measures of preference and report that the participants favored the sans-serif font.

In addition, Boyarski, Neuwirth, Forlizzi, and Regli (1998) measured how quickly participants read and comprehended text on a computer screen and found no effects of typeface. Bernard, Mills, Peterson, and Storrer (2001) assessed "effective reading speed" (which was a measure that combined speed and accuracy) and found only minimal effects of typeface on computer screens (Tahoma, a sans-serif font, was read faster than Corsiva, an ornate typeface). However, the participants tended to claim that the sans-serif fonts were more legible. Bernard, Lida, Riley, Hackler, and Janzen (2002) also found no difference in effective reading speed for sans-serif fonts (Arial, Comic, Tahoma, and Verdana) versus serif fonts (Courier New, Georgia, Century Schoolbook, and Times New Roman). However, it is worth noting that the participants in this study read the serif fonts more quickly and reported that Times New Roman (a serif typeface) was most legible. Furthermore, Morrison and Noyes (2003) compared reading performance between Times New Roman and an ornate sans-serif font (Gigi) and found that the serif font was read more quickly and preferred overall.

In short, the empirical literature on fonts is inconsistent. Nevertheless, I think we can make sense of it. For example, consider the following results, reported by Yager, Aquilante, and Plass (1998). They compared how quickly people could read Dutch (serif) and Swiss (sans serif) fonts on a computer monitor (with white letters on black; both of these fonts use proportional spacing). They found that as long as the luminance was high, there was no difference in reading speed. In contrast, when the letters were shown at a very low luminance, the sans-serif font was better. In general, my reading of this literature is that the differences between serif and sans serif are subtle and most often evident only when conditions are not optimal.

The bottom line: Present print at a large enough size (i.e., at least 9 point) and in conditions of sufficient illumination, and any of the standard (not ornate or "fancy") typefaces are acceptable.

How do you decide which font to choose within a category? Part of this decision will be based on taste—yours and your audiences'. Wilson (2001) reports a survey in which most readers said that they preferred sans-serif fonts for the body of text on computer screens. Why might this be? According to Marshall (2001), "in general sans-serif typefaces look modern. Serif typefaces look more old-fashioned or traditional." That said, in my view looking "old fashioned" or "traditional" may have its place. If you want to convey a sense that careful thought has gone into your argument, it's possible that this message is better conveyed by fonts that do not appear trendy and modern (but to my knowledge, this has never been studied).

Use fixed spacing if the label must be small.

In addition to studying typeface, researchers have examined the effects of the spacing between letters (part of the style). Spacing in which narrow letters (e.g., i) take up less room than wide letters (e.g., w) is known as *proportional spacing*; spacing in which each letter takes up the same amount of room is known as *fixed-width spacing*. Helander, Billingsley, and Schurick (1984) report that people read proportionally spaced fonts faster than fixed-width fonts, and Hill (1997) found that participants read Times New Roman (proportionally spaced) faster than Courier New (fixed-width spaced), but she also found that they read Courier New faster than Arial (proportionally spaced). These results are difficult to interpret because typeface and style were confounded.

Moreover, Mansfield, Legge, and Bane (1996) found that the proportionally spaced Times font was actually *more difficult* to read than the fixed-width spaced Courier Bold (when the heights were controlled)—but this finding was obtained when letters were presented at a size that clearly was smaller than optimal (i.e., small enough to slow down reading). And in fact, participants were actually five percent faster with Times than Courier under good viewing conditions. In this study, words printed in the smallest readable size of Times font were twelve percent larger than words presented in the Courier font.

In a particularly well-controlled study, Arditi, Knoblauch, and Grunwald (1990) reported a comparable result (fixed-width better than proportional spacing) when they compared Times fonts where letters were proportionally spaced with those that had fixed spacing. But again, this finding was obtained when the letters were so small as to approach the limits of legibility. Morris, Berry, Hargreaves, and Liarokapis (1991, cited in Yager et al., 1998) replicated this result, and also found that fixed-width spacing was better than proportional spacing at very small sizes, but the benefits of fixed-width spacing disappeared when the letters were presented at twice the size needed to read them. My advice: Present the labels large enough to be easily read, and then allow aesthetics and consistency with other labels in the display to guide your choice of proportional versus fixed-width spacing.

Use common fonts.

Use fonts that are familiar to the reader, which will make them easier to read. Moreover, if you are going to present your display on a computer or in PowerPoint, take into consideration that fact that not all computers are loaded with the same font set. Printers will also replace unrecognized fonts with another. To be safe, use serif fonts Times New Roman (the most common), Garamond, or Century Schoolbook, or sans-serif fonts Arial, Verdana, or Tahoma.

Words in the same label should be close together and typographically similar.

Take advantage of the laws of perceptual grouping to ensure that words in the same label cohere as a unit; they must not group individually with different parts of the display. The appropriate letters, numbers, and words should be relatively close to one another and should be the same in size, color, and brightness. Labels should not be so close to one another that they group improperly, as do the labels in **don't**.

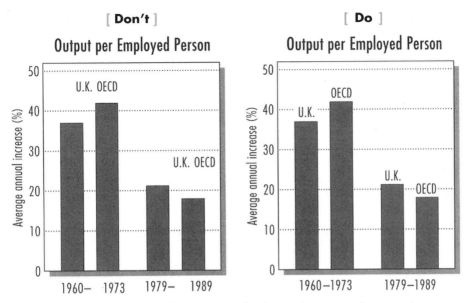

Figure 3.23. If labels group improperly, the reader must work to see what goes with what.

Use the same size and font for labels of corresponding components.

The Principle of Perceptual Organization leads us to group similar forms into units. Furthermore, when we see a difference, we expect it to mean

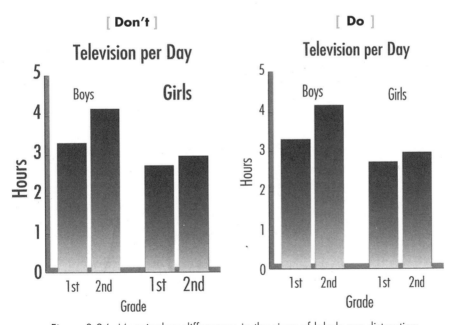

Figure 3.24. Meaningless differences in the sizes of labels are distracting.

something (recall the Principle of Informative Changes) and are confused if it doesn't. Thus, the labels for each of the same type of component (wedges, bars, etc.) should be the same size and font, as should be the labels for the corresponding components of a multipanel display. Similarly, labels on the different axes of line and bar graphs should be the same size and font.

Use more salient labels to label more general components of the display.

The Principle of Salience recognizes that we first notice extreme values and large differences in line length and width, shading, color, and other visual properties. We notice heavier lines and thicker bars before thinner ones, lines that contrast with the color of their backgrounds before lines of a similar color, and brighter colors before dimmer ones. Thus, the label for the display as a whole should be more salient than the labels for any parts, and labels for the individual panels in a multipanel display should be more salient than labels of wedges, objects, or segments.

Use only a few different fonts.

Only change fonts to signal a change in the type of information being conveyed (in accordance with the Principle of Informative Changes).

Do not use underlining.

Underlining cuts off the bottoms of letters, which makes them less discriminable and hence harder to read. Use bold or a change in color for emphasis.

Coordinate color and contrast of words and the background.

Hill (1997) compared the ease of reading many combinations of font and background properties. One important finding was that high contrast between font and background did not always lead to better reading; sometimes lower contrasts were better (e.g., a gray background instead of a white one). This finding buttresses Powell's (1990) advice to avoid using sharp contrasts. Hill also found that she could not state hard-and-fast rules about how to combine colors of font and background (although it was clear that red-on-green is not effective); rather, the precise situation governs the way colors interact.

Use the same terminology in labels and surrounding text.

Using different terms in a display and in the accompanying text suggests that they mean different things (recall the Principle of Informative Changes). Refer to the same animals as "birds" in the graph *and* in the text, not "fowl" just in the text.

Label a dimension with an "unloaded" term.

Many oppositional terms, such as high/low, near/far, above/below, light/dark, and so on, have a marked asymmetry in our use of the words. If you ask, "How high is it?" you are not implying that something is particularly high, but the question "How low is it?" implies that it is low. "High" (which is called the "neutral" or "unmarked" term) names the dimension as well as a pole, whereas "low" names only a pole of the dimension (Clark & Clark, 1977). "Low" is a loaded term—it strongly implies a value. Similarly, "near," "below," and "dark" are loaded terms. When you describe the two ends of the visual continuum, think about the words: Is either loaded? Label the dimension with the term that does not commit you to a particular end of the continuum. In addition, other factors also affect word order, such as the "freezing principle" discussed by Cooper and Ross (1975) and Pinker and Birdsong (1979); according to this principle, the longer, more stressed term in a pair typically is second. In many—but not all—cases the shorter term labels the dimension, so the first word is the neutral one.

Center axis labels and put them parallel to their axes.

Compare **don't** and **do**: In which is it clearer what the axes represent? If you put labels centered and parallel to their axes, the grouping law of common fate will group the label and axis together, as occurs in **do**. If this recommendation cannot be followed because of technical limitations, such as a poorly designed computer graphics program, make sure that the label for an axis is closer to that axis than to the other axis. Avoid putting labels in the lower left, where perceptually they could be grouped with either axis.

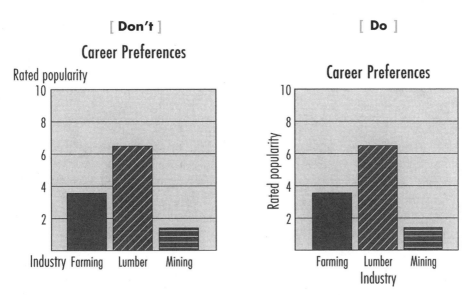

Figure 3.25. Axis labels that do not group clearly with their respective axes are ambiguous and require time to decipher.

Center value labels close to their tick marks.

Center each label of a value along the Y axis—numbers counting off the barrels of oil, tons of peanuts, and so forth—to the left of the appropriate tick mark and closer to it than to anything else, so the label will be associated with the tick. Similarly, center each label of a value along the X axis beneath the appropriate tick mark, and closer to it than to anything else. There is no good reason to place value labels so that they are parallel to the Y axis. Coffey (1961) reports that people can read letters and numbers as easily when they are arranged in a column as when they are arranged in a row. Rotating the numbers will slow readers down (see Jolicoeur, 1990), which is not a good idea unless it is necessary. These guidelines are followed in **do**.

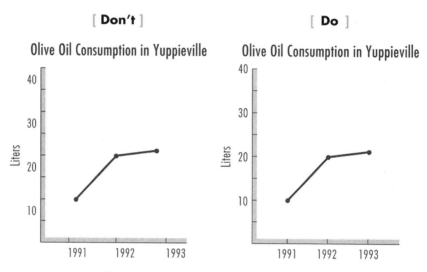

Figure 3.26. If labels are not immediately next to the appropriate tick mark, the reader will have to pause to determine the association.

Label ticks with round numbers and regular intervals.

The Principles of Compatibility, Relevance, and Informative Changes all lead me to recommend round numbers and regular intervals, as in **do**, for tick marks along the axes: 5, 10, 15, 20—not 5.3, 10.4, 15.5, 20.5. The only exception to this rule occurs when a specific value has a special meaning (e.g., 98.6 degrees Fahrenheit). Some computer graphing programs do not respect this recommendation; my advice is to change the graph manually before using it in a written or oral communication.

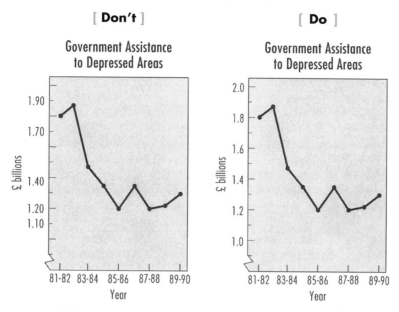

Figure 3.27. Irregular labeling of an axis makes it more difficult to discern values along the content line.

In T-shaped frameworks, label the left and right sets of bars.

This graph, which shows relative percentages of votes for the British Labour (on the left, of course) and Conservative (on the right) parties, is a good use of a side-by-side format. The left and right sets of bars correspond

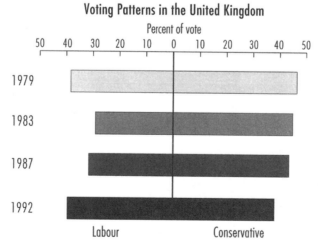

Figure 3.28. Labels for the left and right sides can be at the top or bottom, wherever there is room—provided that proximity properly groups the labels with the content bars.

to different levels of an independent variable. Put labels over or under each side, as space permits.

Do not use redundant labels on the axes of T-shaped frameworks.

If, as is usually the case, the left and right portions of the Y axis delineate the same dependent measure, use one label for both halves (e.g., as done with "Percent of vote" in the previous display). This practice not only reduces clutter but also is consistent with the Principle of Relevance.

Label each component of the content material.

To be easily interpreted, each information-bearing aspect of a display should be labeled, as in **do**. This recommendation follows from the Principles of Relevance and Informative Changes. Recall that "Nutritional Information per Serving" display in chapter 1? It suffered in part from missing labels (specifically, on the arrows that joined the two panels), as does **don't**.

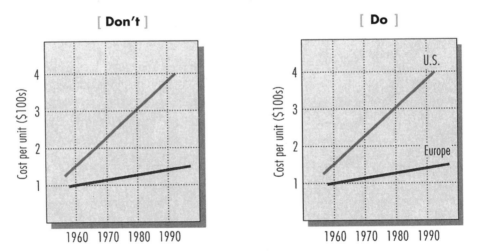

Figure 3.29. Absence of labels can render a display useless.

Label content elements directly.

If possible, place labels within wedges, objects, or segments. If this is not possible because the labels would be too small to be read, then place them as close as possible to the corresponding component, as in **do**, so that they group (because of the grouping law of proximity). If even this is not possible, or creates clutter, then, and only then, use a key.

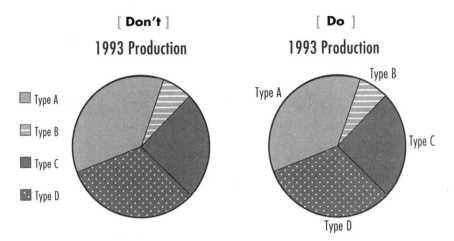

Figure 3.30. Direct labeling greatly reduces the effort required to read a display.

In addition to the results of Milroy and Poulton (1978) summarized in chapter 2, this recommendation is based on results of a study by Parkin (1983, summarized in Pinker, 1990). Parkin investigated the ease of answering questions about the relative heights and slopes of different lines at specific points on the X axis when five different methods of labeling were used: The label was next to the line somewhere along its length (and so was grouped with the line via the grouping law of proximity), was aligned with its end (and so was grouped with the line both via proximity and via good continuation), was aligned with its end but separated by a white space (and so was grouped with the line via good continuation but not proximity), was in a key, or was in a caption. Parkin also varied whether the labels were printed in the same or different color as the line. He found that the participants were fastest when both good continuation and proximity were used to group the labels. However, this result depended on whether the lines formed relatively simple patterns or whether they crossed over many times. When the lines were clearly distinct, putting labels directly next to them was best; when lines formed complex patterns, the three methods that took advantage of grouping laws were equally good. It is of interest that direct labeling was clearly better than use of a key or caption for the simple patterns, but was much less superior for complex (and cluttered) graphs. In general, using the same color for lines and labels also facilitated performance.

Although I have stressed placing labels beneath the X axis, in bar graphs or step graphs, labels can instead sometimes be placed above the bars or steps. If the bars or steps are not arranged in a progression, however, it can be difficult to locate labels above them, and thus I recommend putting the labels beneath the X axis in the usual way. However, if the tops of the content elements are easy to track, as on page 86, then labels are effective when placed directly on the content elements.

Creating the Title

The guidelines for creating labels also apply to the title, and the same additional guidelines for creating the title apply whether the graph stands alone or is part of a multipanel display.

Write a title that indicates what questions are most easily answered by the graph.

A graph will help the reader answer a specific set of questions, and the title should be a signpost to those questions. Which title is more helpful, the one for **don't** or for **do**? When formulating the title, ask yourself, What material is illustrated? Does it apply to a particular time and place? What variable is treated as the "foreground," plotted on the x-axis for line and bar graph variants (except horizontal bar graphs, where it would be on the y-axis)? What are the secondary variables? Mention the foreground variable first, and include only relevant information in the title.

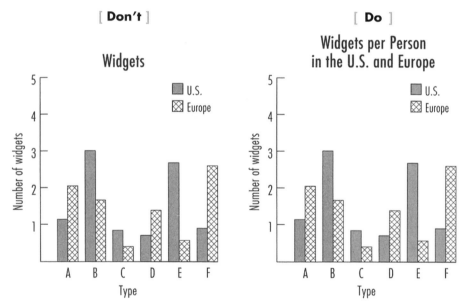

Figure 3.31. The Principle of Relevance will help you formulate the best title for your display.

Ensure that the title is typographically distinct.

Use a larger or different font for the title to help to set it off (exploiting the Principle of Perceptual Organization). Furthermore, ensure that this font is sufficiently distinct to catch the eye immediately, making the title the most salient element of the display.

Center the title over or under the display.

Convention leads the reader to expect the title in one of these two places. If content labels or elements appear at the bottom of the display, it is better to put the title at the top. If you put the title at the bottom, it may be close enough to the other labels or elements to be improperly grouped. If the title is off center, it may be grouped with an axis, as in **don't**. If, however, the display appears in a book or lengthy article with a distinctive but homogeneous design (e.g., one in which the titles are always in the left margin), the reader will grow familiar with this design and will group the title properly. Note that such departures from the norm initially require the reader to work harder than do the standard arrangements.

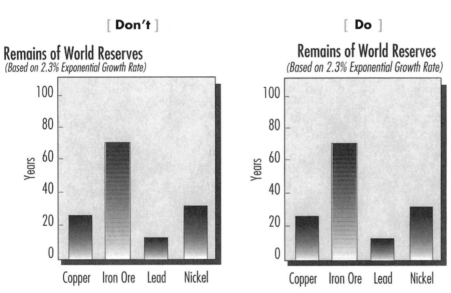

Figure 3.32. The title should clearly be seen to apply to the display as a whole, not an axis or other component.

The Next Step

If you chose in chapter 2 to make a pie graph, divided-bar graph, or visual table, you have used this chapter only to create the title and content labels. You should now turn to chapter 7 to consider color and hatching and other optional elements and then check chapter 8 to be sure that you have not created a misleading graph.

If your choice has an L- or T-shaped framework, you have created and labeled that framework from the recommendations in this chapter and are ready to create the content elements. Depending on the type of graph format you have chosen, turn to chapter 5 or 6.

Creating Pie Graphs, Divided-Bar Graphs, and Visual Tables

4

PIE GRAPHS, visual tables, and divided-bar graphs are the simplest displays that convey relations among quantities. This chapter provides detailed advice for creating the content elements of these graphs. If you want to include shading, hatching, color, three-dimensional effects, inner grid lines, background elements, a key, or a caption, turn to chapter 7 after you have finished with the material in this chapter. Recommendations for placing labels of the content elements are summarized in this chapter;

the reader should turn to chapter 3 for recommendations for creating labels and the title.

Creating a Pie Graph

The content elements of pie graphs are wedges, larger wedges signaling greater amounts. After you have drawn the wedges, you will probably want to fill them in with different shading, hatching, or color, to make each distinct; chapter 7 provides recommendations for these features of displays.

Draw radii from the center of a circle.

The straight lines that define wedges should originate at the precise center of a circle, as in **do**, not at a point off center. The proportion of the depicted amount is reflected by the size of the angle of the wedge. In a graph illustrating the proportions of workers of different ages, if half of all workers are in the 20–30 age range, then that wedge should be 180 degrees—half of the circle; if one quarter of the workers are in the 30–40 age range, then that wedge should be 90 degrees—a quarter of the circle. And so forth.

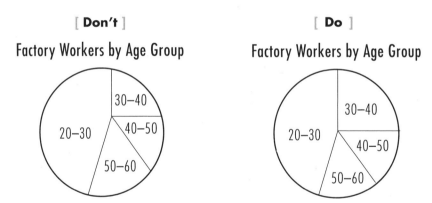

Figure 4.1. Angles, arcs, or chords reflect relative area only if the wedges meet at the center of the circle.

Explode a maximum of twenty-five percent of the wedges.

If you decided to use an exploded pie, you must decide which part or parts to emphasize. The visual system is sensitive to changes in a stimulus. If too many wedges are exploded, as they are in **don't**, they will not be emphasized because the Principle of Salience is being violated.

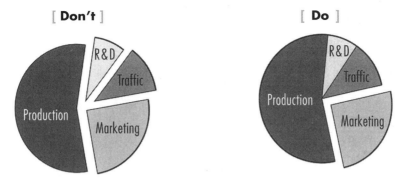

Figure 4.2. If too many wedges are exploded, none stands out.

I offer twenty-five percent as a rough guideline; there is no hard-and-fast percentage. The critical consideration is that enough wedges remain in the pie to make the exploded wedges disrupt the contour of the whole.

Arrange wedges in a simple progression.

Unless you have reasons to order the wedges in a specific way, arrange them in order of size. The Principle of Compatibility suggests that the smallest wedge should be at the top, with size increasing in a clockwise progression. Similar-sized wedges will then be next to each other, as they are in **do**, facilitating comparison of amounts that differ most subtly. Also, the visual system can "chunk" a progression into a unit, making it easier to remember than a set of arbitrarily arranged elements. If people typically read pie graphs by forming chords, it would be important to have wedges that are to be compared at similar orientations: The visual system distorts apparent line length when lines are seen at different orientations. However, as previously noted, Eells (1926) found that only one percent of his participants read pie graphs by estimating the length of chords (fifty-one percent looked at the arcs, twenty-five percent estimated area, and twenty-three percent estimated the angle at the center).

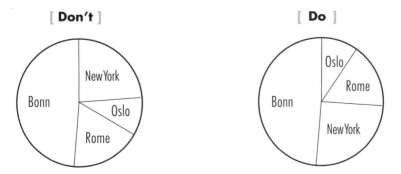

Figure 4.3. In Western culture, readers expect quantities to increase clockwise around a circle.

That said, I must note that Spence and Lewandowsky (1991) found that ordering the wedges as I recommend made no difference in their task. They asked participants to evaluate specific quantities or to compare them, and provided either pie graphs in which wedges were randomly ordered or ordered according to size. However, these are not the only purposes of graphs. For example, there are many studies showing that memory is affected by how material is organized (e.g., see Schacter, 1996). In any event, I again stress that my recommendation must be tempered by the purpose to which the display is put.

It is also of interest to consider the results of a study by Gillan and Callahan (2000). They found that people estimate the proportion depicted by pie segments better when the segments are broken apart and aligned, so that the leftmost portion is vertical in all cases. This procedure essentially converted a pie into a visual table, with the segments being presented separately in a row. This procedure in part helps the reader break the whole into the relevant parts. However, Gillan and Callahan caution that "the aligned pie graph might produce poorer performance in tasks in which wholistic processing of the pie is critical or that explicitly require the graph reader to make part-whole judgments" (p. 588). In addition, an important result from this study is that participants read segments faster when they were vertically aligned than when they were tilted. If so, then if a particular segment is the focus of the message, then you might be better off placing it at the top of the pie.

Place labels in wedges provided that they can be easily read.

The Principle of Perceptual Organization leads to our grouping objects that are physically juxtaposed. Placing labels within the wedges, as in **do**, will clearly associate them with the appropriate wedge. However, do not put labels in wedges if the labels would be too small to read easily.

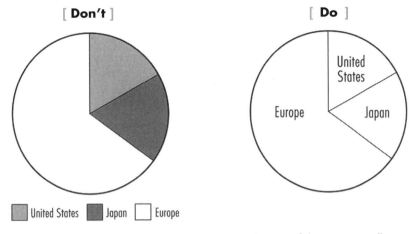

Figure 4.4. Labels group best with content elements if they are actually placed within the elements.

Place labels next to wedges if they cannot all fit within wedges.

If you cannot fit all the labels in wedges, then place all of them next to the appropriate wedges; do not put some labels in wedges, and others next to wedges, as in **don't**. The reader will expect any change to convey information and will otherwise expect consistency (as noted by the Principle of Informative Changes).

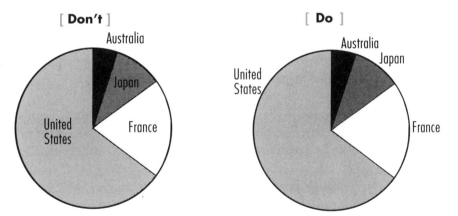

Figure 4.5. A differently placed label can inappropriately signal a change in information.

Creating a Divided-Bar Graph

A divided-bar graph is created by drawing a rectangle—usually vertically oriented—and dividing it into segments.

Make the bar wide enough to demarcate segments clearly.

Because the information in this format is conveyed by the relative sizes of the segments, the bar must be wide enough that the individual segments are seen clearly, as in **do**. Moreover, it is best to make the segments wide enough that labels can fit directly within them, promoting better perceptual grouping.

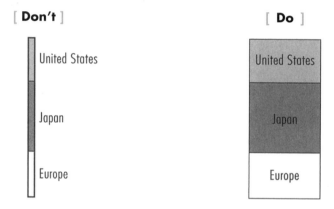

Figure 4.6. Use a bar that is wide enough to ensure that the regions are distinct and to allow the labels to be placed within the content elements.

If you use a scale, place it on the left border.

If you include a scale (always 0–100%), and you have only one bar on your graph, mark the scale directly on the left border of the bar itself, as in **do**. The tick marks should be created and labeled in accordance with the recommendations in chapter 3 (pages 92–95 and 104–105). If you have more than one bar in your display, however, set the scale apart from the bars, at the left. If it were placed directly on the border of the first bar, it would be perceptually grouped with that bar only.

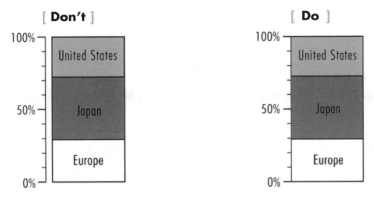

Figure 4.7. A scale on the left border of the bar facilitates lining up tick marks and content segments, but if there is more than one bar a separate scale is preferable, so it does not group with one bar only.

Draw segments with parallel horizontal lines.

The segments should be divided by parallel horizontal lines, as in **do**. If the lines are not parallel, the reader will think that the quantities are varying with time or another variable along an X axis—the divided bar will be mistaken for a layer graph. According to the Principle of Informative Changes, every change in a display will be taken to convey information; the height along the top of a segment should be constant.

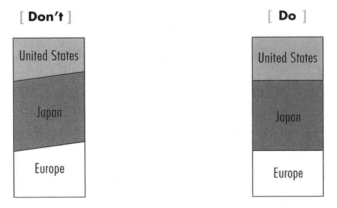

Figure 4.8. Segments drawn with slanting lines improperly suggest a layer graph.

In multiple divided bars, put segments that change the least amount at the bottom.

When designing a divided-bar graph to illustrate two or more independent variables, make it as easy as possible for the reader to compare the changes in the components. Look at **don't**: Were tape sales or singles sales increasing at a faster rate? Now look at **do**: Here it is relatively easy to see that tape sales are increasing at a greater clip. By putting the segment that changes the least amount at the bottom, the comparisons of changing heights will be relatively easy (in order to know the proportion, the reader must visually subtract the bottom of a segment from the top, which is easier when the baselines are similar).

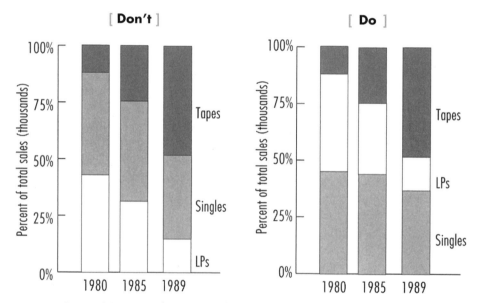

Figure 4.9. Putting the segment that varies least at the bottom of each bar allows the reader to make comparisons among the three categories most easily.

Creating a Visual Table

Visual tables are the least structured type of graph; all that is necessary is that the sizes of content elements reflect relative quantities. Such displays are often interesting because pictures of objects are used as content elements.

Ensure that the appearance of the pattern is compatible with what it symbolizes.

To respect the Principle of Compatibility, the appearance of a pattern should be compatible with what it symbolizes. Our visual systems register the properties of a pattern itself at the same time that we comprehend what the stimulus symbolizes; if the properties of the pattern are not compatible with its meaning, those properties will interfere with our reading the display. The Principle of Compatibility is violated in **don't** because the larger bars actually indicate a *lower* crime rate; this principle is respected in **do** because the larger bars indicate a larger crime rate.

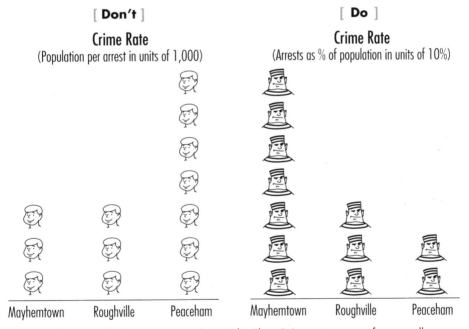

Figure 4.10. Do not convey "more" by "less." A greater extent for a smaller amount violates the reader's expectations.

More generally, visual tables should exploit the fact that we automatically perceive a larger region, a greater extent, a lighter area (on a dark background), or darker area (on a light background) as corresponding to a greater amount: More is more. The visual impression of a difference should correspond to the actual amount of the difference between the represented substances (Brinton, 1919). However, the "more is more" rule must be qualified: Jenks and Knos (1961) report a study of how greater quantities can be represented by increased amounts of ink; they found that the psychological impression of the amount does not increase in equal steps as equal amounts of ink are added. This is an example of the aspect of the Principle of Compatibility concerned with effects of perceptual distortion.

Ensure that pictures used as content elements illustrate the variables.

The Principle of Compatibility implies that pictures used as content elements either should depict the entities being represented, as in **do**, or should have conventional symbolic interpretations that label those entities (e.g., using an icon of an envelope for a post office or a donkey for the Democratic party). If the amount of lumber produced in different regions is illustrated, pictures of trees of different heights would be appropriate, whereas pictures of objects made from trees, such as ladders or park benches, would not.

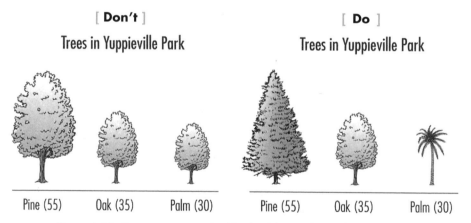

Figure 4.11. Pictures in a visual table should provide another kind of label and should not interfere with the interpretation of the display.

Compare extents at the same orientation.

The **don't** version immediately demonstrates why this recommendation is a good idea: It is much harder to assess and recall the value of each individual element and to compare it with others when there is no common baseline. Moreover, our visual systems systematically misinform us that a vertical line is slightly longer than a horizontal one of equal length (Graham, 1937). The famous top-hat illusion shown in chapter 1 illustrates this distortion nicely (where the hat is actually as wide as it is high, but appears much higher). It has been shown that people tend to overestimate the length of vertical bars, which led Jarvenpaa and Dickson (1988) to advise using horizontal bar graphs if individual data points are compared. Nevertheless, there is no evidence that the distortion of vertical lines seriously affects comparisons among vertical lines (all vertical lines are distorted the same way). In fact, we do not systematically underestimate or overestimate relative differences in lengths at the same orientation. Thus, as long as you stick with a single direction, the lengths of objects in a visual table can convey relative quantities accurately. For additional discussion of illusions, see Frisby (1980), Gregory (1966, 1970), and Robinson (1972).

[**Don't**]

U.S. Exports of Manufactured Goods

[**Do**]

U.S. Exports of Manufactured Goods

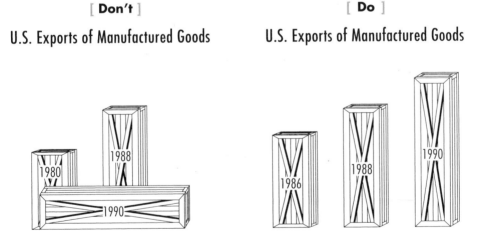

Figure 4.12. It is much easier to compare extents at the same orientation that at different orientations.

In general, do not covary height and width of objects to illustrate amounts.

There are two ways that we perceive combinations of visual properties: First, we register separable properties individually; a reader can easily pay attention to one variable and ignore the other. The color and length of a line are separable, as are the length and orientation of a line. Second, some properties are automatically combined by the visual system, and the Principle of Perceptual Organization (specifically, the aspect of integrated vs. separated dimensions) states that we do not register these properties separately except with difficulty. The height and width of a rectangle are examples of integral dimensions: You cannot easily pay attention to the height without also registering the width and the resulting area—and we do not perceive differences in area very accurately. For this reason, Macdonald-Ross (1977) recommends against varying the overall size of a picture—which varies both height and width—to convey information; instead, he recommends varying the number of pictures (each one standing for the same fixed amount) to produce a row or column, the length of which specifies the total value (as in Figure 4.10).

Do not vary height and width or other integral dimensions to specify separate variables.

A designer might use rectangular objects to present seasonal totals for new accounts opened at a bank. The height of each rectangle would indicate the number of accounts that were opened during that particular season; the width, the average amount deposited in such accounts; and the third variable, the area, would indicate the total revenue for each season. This scheme is shown in **don't**. One problem with such a display is that we see a rectangle—not height, width, and area as distinct elements. Height and width are integral dimensions, as just discussed.

In general, do not vary two or more integral dimensions to convey different sorts of information. Garner (1970, 1974, 1976) suggests two criteria for determining whether two dimensions are integral: difficulty in ignoring one dimension while attending to the other, and facilitation in reading one dimension when the value on the other provides redundant information. A weaker form of this relation occurs with "configural dimensions," which exhibit only filtering interference. For example, the angle formed by two lines is configural with the slopes of the lines (see Beringer, 1987; Carswell & Wickens, 1987a, 1987b, 1988, 1990; Clement, 1978; Jacob, Egeth, & Bevan, 1976; Pomerantz, 1981).

Nevertheless, Carswell and Wickens (1990) discuss ways in which one might try to pack more information into a display using integral dimensions, but these displays are useful only in limited circumstances, when the reader must attend to—and integrate across—several dimensions to make a deci-

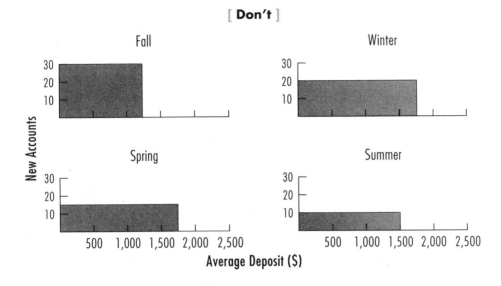

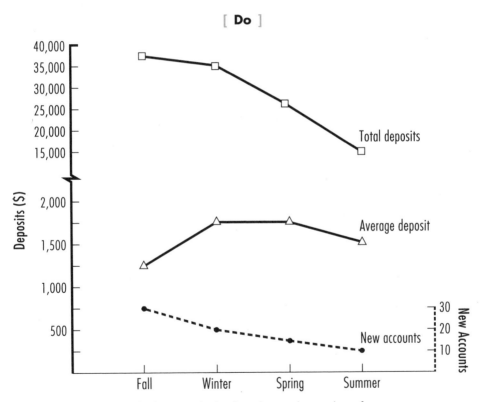

Figure 4.13. The line graph clearly indicates the number of new accounts, the average deposit, and the total deposits per season, as well as the relations among these measures. The same information is embedded in the four panels above, but considerable effort is required to decode it—we perceive height and width integrally.

sion (see also Barnett & Wickens, 1988; Carswell & Wickens, 1987a, 1987b). Indeed, Casey and Wickens (1986) and Jones and Wickens (1986)—both cited in Carswell and Wickens (1990)—found that integral displays produced no better performance, and sometimes worse performance, than nonintegral displays even when the participants had to process correlated variables (see also Coury, Boulette, & Smith, 1989; Sanderson, Flach, Buttigieg, & Casey, 1989). Moreover, it seems clear that the efficacy of integral or configural displays depends critically on how variables are assigned to specific dimensions or features of the display (MacGregor & Slovic, 1986).

An oft-cited use of configural information in displays was described by Chernoff (1973; see also Chernoff & Rizvi, 1975). Chernoff designed schematic faces in which the shape of each feature conveys information about a separate variable; the overall expression of a face was intended to convey an impression about the relations among the variables. Jacob, Egeth, and Bevan (1976) and Wainer and Thissen (1981) note that people can evaluate the psychological "distance" between the faces. However, these displays are tricky—the features do not appear to be equally important in creating the overall impression (cf. MacGregor & Slovic, 1986) and sometimes may be difficult to read. Numerous variants have been proposed (e.g., Wakimoto, 1977; Wainer, 1979; Wainer & Thissen, 1981).

In short, although using integral and configural dimensions is sometimes an effective way to pack much information into a single display (e.g., Bickel, Hammel, & O'Connell, 1975), readers typically need considerable practice before being able to decode such displays easily.

The Next Step
Consult the recommendations in chapter 7 for using shading, hatching, color, three-dimensional effects, background elements, and keys. Turn to chapter 3 to create the labels for the individual content elements and the display as a whole.

Creating Bar-Graph Variants

5

Bar graphs and their variants are used primarily to answer questions about quantities: how they compare and how they change. Bar-graph variants must include an L-shape or T-shape framework as well as content and labels. (Unlike bar-graph variants, divided bars need not include a framework and always display proportions—and thus they are closer to pie graphs than to the kinds of displays considered in this chapter, and are discussed in chapter 4.) Create the framework and its labels using the recommendations in chapter 3. Then create the content and place its labels using this chapter. When using and producing labels of content elements, follow the guidelines and recommendations illustrated and discussed in chapter 3. If shading, hatching, color, three-dimensional effects, an inner grid, a key, background elements, a caption, and/or more than one panel is needed, consult the recommendations in chapter 7 after finishing this chapter.

Bar Graphs

A bar is any content shape whose extent (height in vertical bar graphs, length in horizontal bar graphs) is used to indicate an amount; rows of pictures—houses, trees, people, or other images—can be used as bars within a framework.

Do not insist that bars minimize "ink."

Some of the most interesting research on how to construct bar and line graphs has focused on evaluating Tufte's (1983) admonition to "Maximize the data-ink ratio, within reason" (p. 96). The data:ink ratio is the "proportion of a graphic's ink devoted to the non-redundant display of data" (p. 93). A corollary of this rule is that one should eliminate redundant ink used to convey data and all ink that does not convey data. For example, a bar should be drawn simply as a vertical line with a short horizontal segment at its top (representing one side and the top of a bar—eliminating the redundant ink of the other side of the bar and the internal filling). The intended consequence of such editing is to increase the salience of the data-conveying elements of a graph. However, the key caveat in Tufte's recommendation is the qualifier, "within reason." Carswell (1992a) analyzed the results of 39 experiments in which the effects of this variable could be examined, and did not find support for Tufte's principle.

Nevertheless, Gillan and Richman (1994) did provide evidence that under some circumstances following Tufte's principle does facilitate graph comprehension. Specifically, they showed participants three versions of line and bar graphs, which varied in their data:ink ratios—from very low (which included detailed backgrounds, which unfortunately confounded the distraction of very salient background objects with amount of nondata ink), to intermediate (standard line and bar graphs), to high (line graphs with no axes or lines connecting dots that indicated data points and bar graphs with no axes and only the horizontal segment at the top to indicate bar height). All of the stimulus graphs displayed only two data points. The participants were asked to compare the two data points, to compute a difference between them, or to compute a mean value.

Gillan and Richman (1994) found that in some circumstances the participants responded more quickly with the pared-down graphs. For example, these researchers found that participants did better when there was no background that had visual features similar to the data-conveying elements (bars or graphs), and that participants did not benefit from tick marks on the Y axis (but the axis still had the values labeled) when they took differences or means.

However, in general, the findings indicated that the data:ink ratio had different effects for different types of displays and tasks. For example, if anything, participants took a bit more time to compare values with the pared-down bar graphs than with the standard ones, and took a bit more time to assess the mean for the pared-down line graphs than the standard ones. Gillan and Richman note that "ink can be helpful or detrimental to a graph reader depending on the function and location of the ink, the user's task, the type of graph, and the physical relations among graphical elements" (p. 637).

For example, the participants did not generally do better when the lines representing the X and Y axes were removed, but eliminating the vertical line (for the Y axis), while leaving intact the value markings along it, did not affect performance with bar graphs as much as with line graphs. The results, taken as whole, indicated that interactions among different features of a display were more important than any individual feature in its own right. In the words of the authors, "Thus, these data call into question Tufte's general rule that graph designers should maximize the data-ink ratio by eliminating non-data-ink and redundant ink" (p. 638).

Similarly, considering his results in toto, Siegrist (1996) comments:

> These results can be used to evaluate Tufte's (1983) hypothesis that charts with a high data ink ratio are superior to charts with a low data ink ratio. This hypothesis is not supported by our data. We therefore agree with Spence (1990) and Kosslyn (1985), who had some doubt as to the wisdom of Tufte's recommendation for empirical and theoretical reasons. (p. 100)

Mark corresponding bars in the same way.

Try to compare the performance of the two brands of Coca-Cola in 1989 from the **don't** graph. This is not as easy as it should be because in a misplaced attempt at variety in the way the bars are shaded, the designer used different shadings for corresponding elements. When more than one independent variable appears in a bar graph (as, here, brand and year), use different colors, shades, and/or different dashed, striped, or hatching patterns to indicate the bars for different variables. The corresponding bars in each cluster should be marked the same way so that they are grouped appropriately (exploiting the grouping law of similarity), as in **do**.

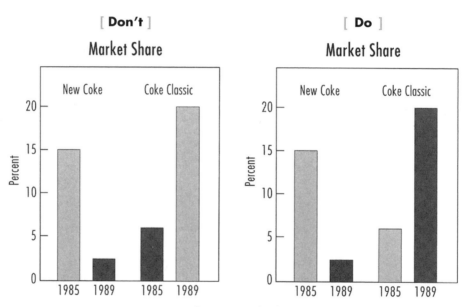

Figure 5.1. Inconsistent marking groups bars improperly in **Don't**, making it harder to compare corresponding elements.

Arrange corresponding bars in the same way.

Arrange bars that depict the levels of a parameter in the same way in each cluster on the X axis. Say you want to graph numbers of voters in two counties, considering separately young and old men and women. County would be on the X axis, and four bars, one for each of the four demographic groups, would sit over each county. The demographic groups—young men, old men, young women, old women—could be indicated by different shading. The order of the bars for voter group should be the same for each county, as in **do**. If it isn't, the Principle of Informative Changes implies that the reader will waste time trying to make comparisons and wondering why the order differs.

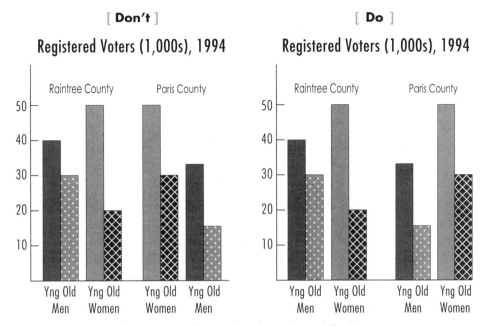

Figure 5.2. A gratuitous change of order makes it difficult to compare corresponding bars.

Ensure that overlapping bars do not look like stacked bars.

Some bar graphs show two parameters by partially overlapping one of the two bars in each pair, so that one appears to be standing in front of the

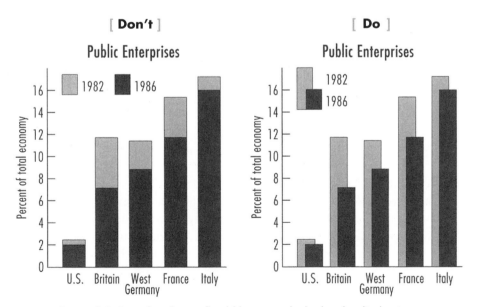

Figure 5.3. Partial occlusion should leave no doubt that the display is not a stacked-bar graph.

other. If you do this, make sure that the "nearer" bar does not cover all or nearly all of the lower part of the "farther" one. If it does, your display will be misinterpreted as a stacked-bar graph, and your readers confused. At first glance, **don't** seems to indicate the sum of the percentage of public enterprises in the total economy for 1982 and 1986. In fact, as is evident in **do**, the two bars in each pair specify entirely separate measures.

Do not vary the salience of individual bars arbitrarily.

Unless the emphasis is intentional, no bar should stand out from the others as happens in **do**; if it does, the difference in salience will lead the reader to notice it first and assume that it is more important than the other bars in the display.

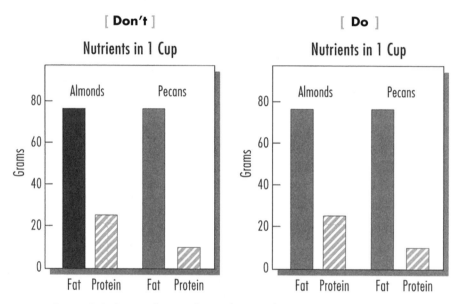

Figure 5.4. Great salience of one element, if not intentional, gives it an undesirable role as the center or attention.

Leave space between bar clusters.

Is it easier to compare the market shares in 1985 for the three brands in **don't** or **do**? When two or more independent variables are included in a bar graph, group the bar clusters over each appropriate location on the X axis and leave extra space between the clusters. As a rule of thumb, the space between clusters should be noticeably wider than a bar. The grouping law of proximity will group the bars within a cluster but separate those in different clusters, as can be seen in **do**.

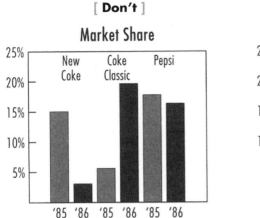
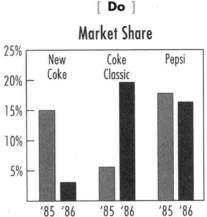

Figure 5.5. Six perceptual units are too many to apprehend immediately; proper grouping not only makes the display easier to take in but also helps to group the bars with their labels.

Ensure that bars generally do not extend beyond the end of the scale.

Bars should not extend over the top of the Y axis as in **don't** (or to the right of the X axis, in a horizontal bar graph) if the viewer is supposed to be able to extract specific point values; mentally continuing the axis and its scale requires effort, and the Principle of Relevance requires that necessary information be supplied. However, if the point of the display is not to show specific values but only to indicate that, say, housing prices have gone through the roof, you should not present extraneous details (such as tick marks, labels on the Y axis, etc.). Some of the most effective graphs in *Time*

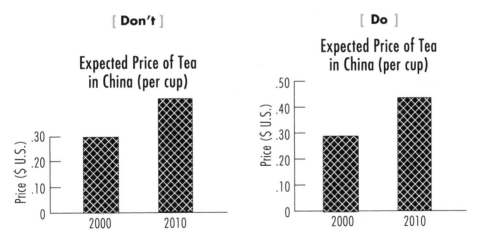

Figure 5.6. If the Y axis is too short, the reader cannot immediately estimate the price of tea in 2010.

magazine have tall bars that extend into the text, driving home the point that some trend has exploded beyond its usual boundaries.

Use half "I" error bars.

Error bars specify a range of measurement error around an average. For example, in a graph of the results of an opinion poll, error bars would indicate the range over which one could expect to find the average if a different sample had been polled at the same time. In a bar graph, show only the half of the "I" that extends above the bar, as illustrated in **do**. The error bars should be drawn in lines of the same thickness as those used for the framework, which will help prevent them from being overshadowed by more salient elements of the display.

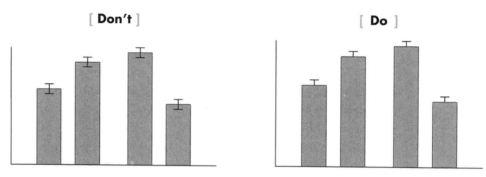

Figure 5.7. The half-"I" bars convey the necessary information without cluttering the display.

Use hierarchical labeling.

Which town had greater sales of single-family homes in 1991, Hillsborough or Pinellas? This is relatively easy to determine from **do**, but much harder from **don't**. When there are two or more independent variables (and so there is at least one parameter), use a hierarchical labeling system, one in which some labels are larger or bolder—that is, more important—than others. Such a scheme, illustrated in **do**, eliminates the need for a key, which would tax our short-term memory capacities. It also avoids redundant labeling, which clutters a display (thereby making it more difficult to discriminate the other elements) and specifies each of the relevant dimensions separately—which helps the reader to make specific comparisons.

By using labels of greater salience to label larger portions of the display, you draw the reader's attention to major labels before minor ones; in order to understand the scale values, the reader must know what variable is being graphed on that scale, and in order to understand why those variables are presented, the reader must understand what is being presented.

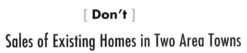

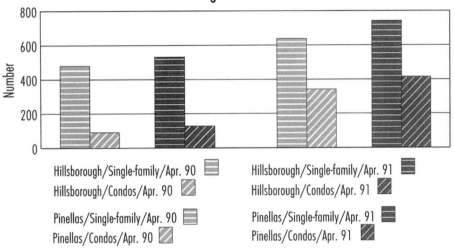

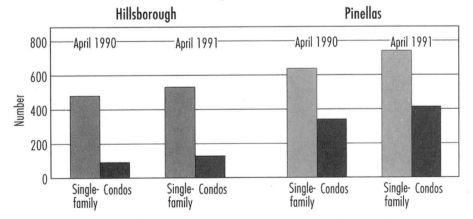

Figure 5.8. The bars are easier to compare along specific dimension (dwelling type, town, year) when the labels specify the dimensions separately.

Stacked-Bar Graphs

Stacked-bar graphs are a hybrid of bar graphs and divided-bar graphs. They are like bar graphs in that the bars appear in an L-shaped framework, and they are like divided-bar graphs in that they display components of a whole. However, in stacked-bar graphs, the components do not sum to a constant-sized whole.

Put the segments that change the least amount at the bottom.

Like divided-bar graphs, stacked-bar graphs work best when you put the segment that changes the least amount at the bottom, as in **do**; this practice makes it relatively easy to compare the changing amounts of specific segments. To read a stacked-bar graph, the reader must visually subtract the value at the bottom of a segment from the value at the top to see the amount, a feat that is more easily accomplished if the bottoms are as nearly at the same heights as the data allow.

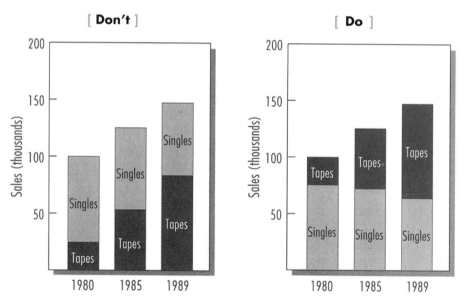

Figure 5.9. Because "Singles" changes relatively little, putting it at the bottom provides a relatively constant baseline for "Tapes," which helps the reader to compare the two categories.

Step Graphs

Step graphs are bar graphs in which the bars are contiguous and the internal borders have been removed, resulting in a step pattern. The content element of a step graph is a line—the outline of the steps—that varies in height in discrete increments. This format is best for displaying a single independent variable. Because there is only one independent variable, which is labeled directly or along the X axis, there are no additional labels for the content material.

Ensure that the line is detectable.

The content line should be heavy enough to be noticed at a glance, as in **do**. The content lines should be at least as heavy as the framework; indeed, most of the graphs in this book—the **do** versions, that is—show content lines three times heavier than the framework.

[**Don't**] [**Do**]

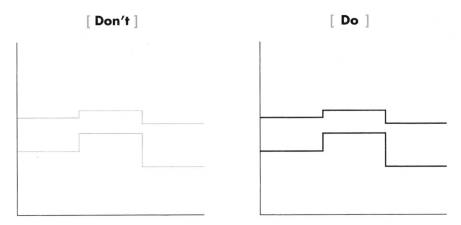

Figure 5.10. The reader should not have to search for the content of a graph.

Make the steps of equal width.

In which version of the graph does U.S. investment in Germany look more important? Because of the Principle of Perceptual Organization (the aspect concerned with integrated versus separated dimensions), the reader will perceive a wider step as representing a greater amount, even if that step

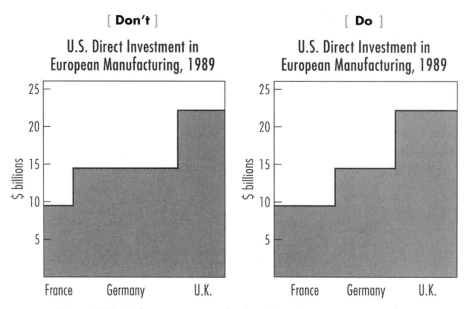

Figure 5.11. Making one step wider than the others gives it unequal importance.

is the same height as another, narrower step. Make sure the steps are of equal width, not drawn as in **don't**.

Fill the area under the line with a single pattern or color.

The primary virtue of step graphs is that they produce patterns that indicate specific trends. As is evident in **do**, if the area under the line is filled

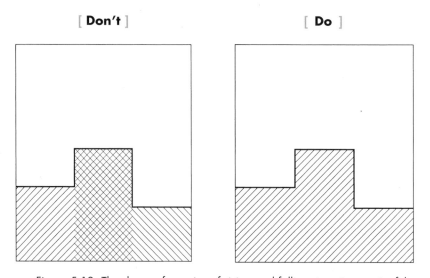

Figure 5.12. The shape of a series of rising and falling steps is meaningful and should not be disrupted.

with a color or texture that is different from the background, the grouping law of similarity will lead the reader to see that area as a single shape, facilitating recognition of a trend. Only a single color or texture should be used; otherwise, the impression of a single region will be disrupted.

If the area under the line is filled with a color or texture different from that of the background, the salience of the content will be increased and the eye will be drawn to this visual difference. If the reader is likely to be familiar with the topic of the display (say, from the surrounding text), this is fine. However, in some situations you may want to ensure that the reader examines other components of the display, such as the title, carefully; in these cases, consider whether the content material should be more salient than the framework and labels.

Side-by-Side Graphs

In many respects, side-by-side graphs are like horizontal bar graphs; they differ in the type of framework used.

Align the bottoms of corresponding bars.

The power of side-by-side graphs arises from the perceptual units that are created by pairs of corresponding bars. Therefore, the bars in each pair must be aligned so that the same extent corresponds to the same amount, as in **do**. The central vertical axis can produce an asymmetric pattern, allowing the reader to see the inequalities in the bars immediately.

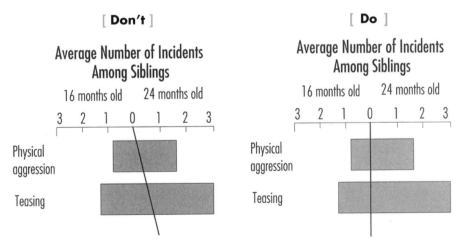

Figure 5.13. The slanted Y axis shifts the starting points of the bars, resulting in a distorted impression of their relative magnitudes.

Order the bars to emphasize interactions.

The visual system is sensitive to differences in patterns. Thus, if the Y axis is a nominal scale and the levels have no inherent order, arrange the left series of bars to form a regular progression, as in **do**; the differences between those bars and the ones on the right will then be readily apparent.

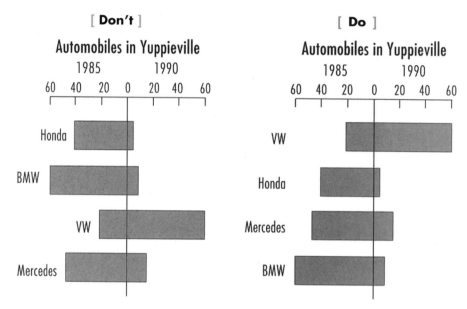

Figure 5.14. A smooth progression of the left-hand bars provides a good contrast to the pattern of the right-hand bars.

Label pairs of bars consistently on one side.

A label for each pair of bars should be placed either to the left or right of the pair, close enough to be grouped both by good continuation and proximity. The Principle of Informative Changes implies that the labels should be in the same places for each pair. As a general rule (at least for Western readers), place labels on the left, following the Western convention of reading left to right

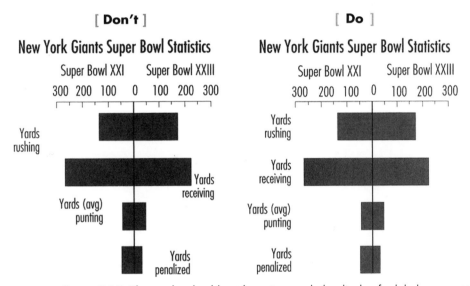

Figure 5.15. The reader should not have to search the display for labels.

The Next Step

Now that you have created your graph in outline, the next step is to fill in the bars with shading, hatching, or color. You may need to add a key. Other possibilities are three-dimensional effects, an inner grid, a caption, background elements, and a multipanel display. Turn to chapter 7, where you will find recommendations for creating all of these features.

Creating Line-Graph Variants and Scatterplots

6

Line graphs, layer graphs, and scatterplots all have an L-shaped framework, and recommendations for producing the framework (and labeling and titling it) were given in chapter 3. In addition, to use and produce labels of content elements effectively, follow the guidelines and recommendations illustrated and discussed in chapter 3. The material in the following pages will guide you, step-by-step, to create the content of line-graph variants and scatterplots.

Line Graphs

Line graphs, because of the continuous nature of their content, are particularly useful in showing how quantities change over time.

Vary the salience of lines to indicate relative importance.

The most salient line will be noticed first and interpreted as the most important (this is the Principle of Salience at work). If you are looking at all three networks' morning shows and comparing their ratings, the graph on the left is the style you want. But if your subject is NBC's *Today* show and you want to emphasize that show's ratings against the other two shows in the field, the one on the right better illustrates your focus.

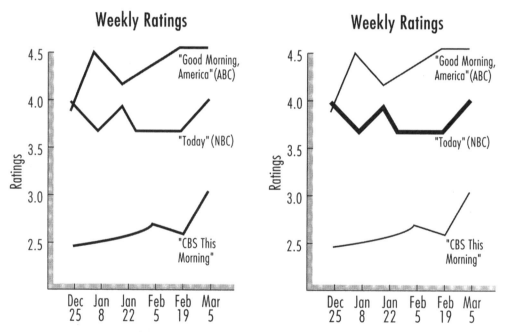

Figure 6.1. If the focus is on NBC, the graph on the right is preferable; the increased salience of the line for NBC immediately draws the reader's attention.

Ensure that crossing or nearby lines are discriminable.

When lines are nearby or cross often, special care must be taken to ensure that they are discriminable. One way to increase discriminability is to use different dashed lines; another way is to use different colors (see chapter 7).

Make dashes in lines differ by at least 2 to 1.

To be immediately discriminable from one another, dashed lines should differ in elements per inch in a ratio of at least 2 to 1. For example, if one line has four dashes to the inch, no other line should have more than two or fewer than eight dashes to the inch. This recommendation is based on the Principle of Perceptual Organization, specifically, the aspect concerning input channels. The visual system registers information at different levels of scale (see pages 8–10). If the differences in levels are roughly 2 to 1, they will be processed by different "input channels"; variations less than that will be processed by the same channel, and the distinctions between them perceived only with effort. If you have so many lines that it is not possible to maintain a 2-to-1 ratio of elements per inch, use lines like those illustrated in **do**; it has been shown that these lines are perceptually highly discriminable (Schutz, 1961b).

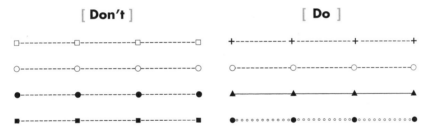

Figure 6.2. Studies have shown that the four lines and symbols on the right are highly discriminable.

**If lines connect discrete points, make the points
at least twice as thick as the line.**

Some line graphs connect discrete points that mark specific levels on the X axis. These points should be specified by dots or symbols, which should be at least twice as thick as the line. Again, the effects of input channels are evident: 2-to-1 ratios of size (dot-to-line) are desirable when relatively

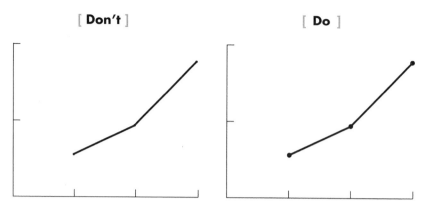

Figure 6.3. The points on the lines are especially important and should be emphasized by discriminable dots.

small marks must be distinguished immediately. The **do** version follows this recommendation.

Use discriminable symbols for points connected by different lines.

When lines are nearby or cross often, discriminability can be enhanced by using visually distinct symbols for important points—either in conjunction with different dashed or colored lines or with solid lines. The symbols used must be as discriminable as possible. Plus signs, Xs, open circles, and filled triangles remain distinct even when reduced to small sizes. The difference between filled and unfilled version of the same element, however, often is difficult to discern after photoreduction.

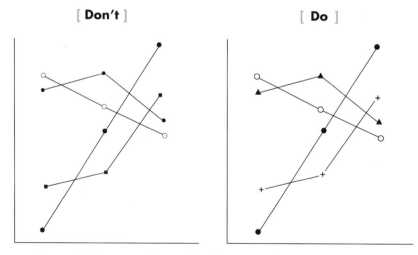

Figure 6.4. Discriminable symbols help the reader to keep track of lines that are nearby or that cross often.

Do not fill in the area between two lines to emphasize relative trends.

At first glance, it might seem like a good idea to fill in the area between two lines in a line graph, thereby creating a shape, as in **don't**. In accordance with the Principle of Compatibility, features of the shape itself will convey information about the relationship between the two functions: Shapes that are larger on one side than the other tell an instant story about the relative changes in the two levels of the parameter. Also, the large distinctive area will be salient, drawing the reader's attention to this relation. There are two potential major drawbacks to this practice, however. First, the filled-in area might be salient to such a degree that the reader will ignore other lines in the display. Second, the reader might mistake a line graph for a layer graph and attempt to read a cumulative total, as one would in a layer graph, not the absolute independent values indicated by two or more lines on a line graph.

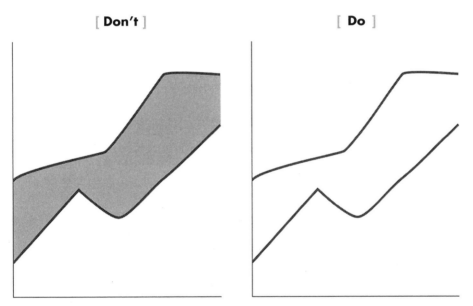

[Don't]   **[Do]**

Figure 6.5. The filled-in region is not only distracting, it may also mislead the reader into thinking this is a layer graph.

Ensure that error bars are discriminable by using half-"I" bars.

The error variation around a mean (i.e., the arithmetic average) should be illustrated with a small "I" bar on the dot that represents the mean. The distance above and below the dot indicates the likely range of variation if the data were collected again in the same circumstances (see appendix 1). If the "I" is centered on the point, it is grouped with it by the grouping laws of proximity and good form. If error bars happen to overlap, include only the top half of the uppermost bar and the bottom half of the lowermost bar, as illustrated in **do**. This way the bars will be more discriminable, the display less cluttered. The degree of overlap is part of the message, providing information about the ranges of variation, but the reader should be able to tell which error bars belong to which points. In addition, in some cases the error bars are so large that they threaten to overwhelm the trend data. This may be appropriate if the error is so large that the graphed differences are not significant, but if the visible differences in mean values are to be taken seriously, you might want to make the lines connecting the dots more salient than the lines used to draw the error bars. This recommendation has been followed in the example below.

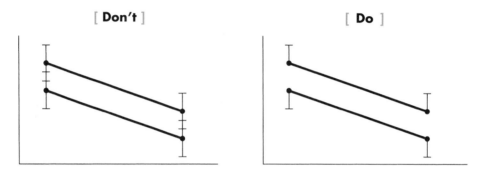

Figure 6.6. Use the technique illustrated at the right to make error bars discriminable, unless your message is that variability is so large that points are not really distinct (which is unlikely if you want to graph them).

In mixed line and bar displays, make one function primary.

If two types of data are strongly related, you might want to include them in the same display even if a different format is appropriate for each type—but, as noted in chapter 2, I do not recommend using mixed line–bar displays if your goal is to illustrate an interaction. Even if you do not want to show an interaction, such displays are apt to be complex; thus, these displays are most effective if one component is made secondary to the other, leading the reader to take in the display one part at a time.

For example, "Business Registrations and Failures" includes a line to indicate the number of new businesses (one dependent variable) in each of several years (the independent variable), and a set of bars to indicate the number of businesses that went under (another dependent variable) in each of those years. The designer wanted to emphasize that the number of new businesses was steadily climbing and so used a line graph for that variable, while the bar graph component allows the reader to read off point values for the number that failed. Because the editorial emphasis is on the positive aspect of the data, the number of startups is visually emphasized. As in this example, the more important dependent measure—in this case, new businesses—should be presented on the left Y axis, and the corresponding data should be plotted near the top; because we read from left to right and top to bottom, information in these positions is most likely to be attended to first.

Furthermore, it is useful to make the more important information more salient. The scales for the dependent measures can be adjusted so that one function can be placed above the other, either by choosing appropriate distances for the tick marks or by excising part of the axis (see pages 86–91 and 93–95).

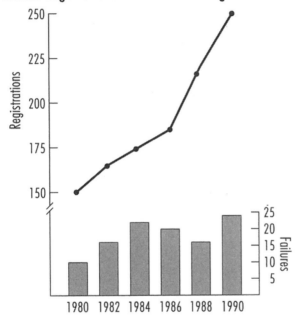

Business Registrations and Failures in England and Wales

Figure 6.7. The heavy black line, drawn in the upper portion of the display, draws the reader's attention to business registrations, the main topic of the display.

Put labels of all lines in the same part of the display.

Look at **don't** and decide which group is growing more quickly, Liberals or Others. This is more difficult to determine than it has to be because of the way the labels are placed. Whereas **do** uses the laws of perceptual grouping to pair labels and lines, **don't** is a victim of these laws. Labels of different lines should line up, forming a column; otherwise, the viewer will have to search for the different labels in different areas of the graph. In accordance with the Principle of Informative Changes, readers expect not to see changes unless they carry information. Pick an area where the lines are

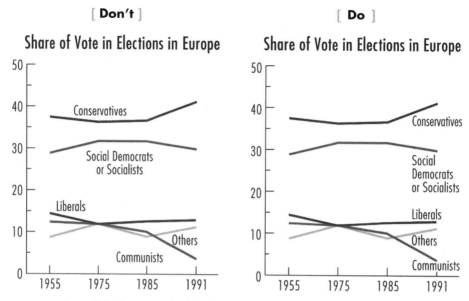

Figure 6.8. Searching for labels is time-consuming; the arrangement on the right is not only more efficient, but also less busy visually.

least cluttered (recall the Principle of Discriminability), and put all labels in corresponding positions.

Position labels at the ends of lines.

The previous recommendation often implies this one; the only reason not to label lines at their ends is if the labels are too large to fit into the available space. Putting a label at the end of each line is a good practice because the grouping law of good continuation will serve to group the labels with the lines.

Use hierarchical labeling.

A hierarchical labeling scheme using different levels of labels minimizes the effort required of the reader. Look at **do**: Are sales of both condos and single-family homes increasing more in one community than the other? The hierarchical labeling scheme allows the reader to compare the data along individual dimensions (dwelling type, location) and eliminates redundancy and its attendant clutter. If the lines are not arranged to permit hierarchical labeling, either fully label each line directly with the name of the independent variable and level (keeping the order of the two names constant) or—

if this produces a cluttered display—use a key. If each information-bearing element is to be labeled, these are the only choices.

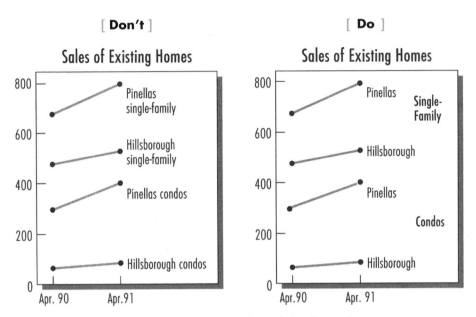

Figure 6.9. A hierarchical labeling scheme helps the reader to compare data along one dimension (here, dwelling type or location) and reduces clutter.

Label critical point values explicitly.

The primary purpose of line graphs is to illustrate specific trends or interactions among values. If you also want to emphasize a few specific point values, those values can be labeled directly (Culbertson & Powers, 1959), as in **do**. However, respecting the Principles of Relevance and Capacity Limitations, it is best not to label specific point values unless they are particularly important. For example, if the display is supposed to emphasize the peak and closing value real estate prices during a specific decade, and the precise amounts of the absolute high and closing values are important (rather than simply the visual impression of their relation), put only those two numbers on the display. The numbers should be close to the appropriate parts of the line (so they are grouped in accordance with the grouping law of proximity), and the title should indicate what the numbers mean.

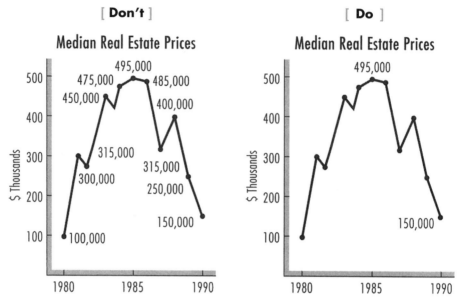

Figure 6.10. A graph that provides more information than readers need forces them to filter and search. Directly label only the critical values.

Layer Graphs

A layer graph is like a set of very narrow stacked bars that are pushed together. These displays illustrate changes in components of a whole over time or another interval scale along the X axis.

Because layer graphs are a hybrid of line graphs and stacked-bar graphs, most of the recommendations offered here follow from recommendations made elsewhere for these other formats. Any possible confusion between a line graph and a layer graph can be eliminated by proper titles, labels, and captions.

Layer graphs differ from stacked-bar graphs in only two important ways: The X axis must specify an interval scale, and there are no vertical lines demarcating discrete intervals for levels along the X axis. The content elements are lines that define filled-in regions.

Use shading to ensure that a layer graph is not misread as a line graph.

Without shading, it is not immediately clear whether **don't** is a line graph or a layer graph—whether the lines represent independent trends considered as parallel events (line graph) or boundaries of areas that sum to a total represented by the height of the top line (layer graph). Any doubt is removed in **do**. However, when using shading, you must ensure that the labels are discriminable, which can be a challenge (as shown in **do**) if you need to use relatively dark shading. By the same token, use patterned backgrounds behind type with caution.

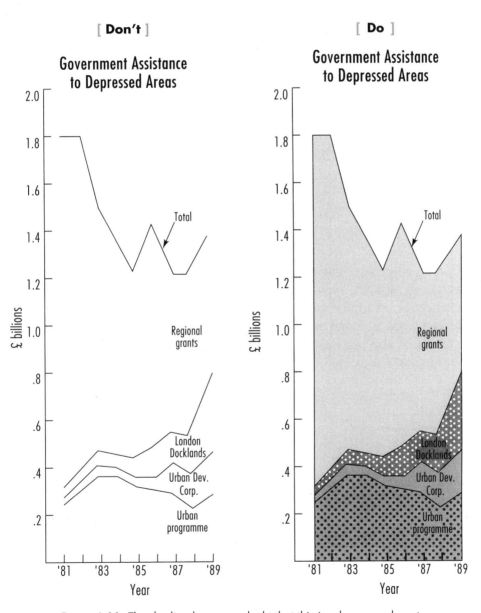

Figure 6.11. The shading leaves no doubt that this is a layer graph, not a line graph.

Put the layers that change least at the bottom.

By putting the layers that change the least at the bottom, you can make it easier for the reader to compare the degree of change in specific segments. This recommendation is analogous to those illustrated on pages 118 and 134. In general, any time you stack or layer quantities, put what changes least on the bottom.

Scatterplots

Scatterplots employ dots and symbols to represent discrete points, each of which typically corresponds to a single measurement. These displays are useful for spotting trends in data. In most respects, scatterplots are like line graphs without the lines; hence, the recommendations that pertain to line graphs—other than those that apply to the line itself—also apply here.

Ensure that point symbols are discriminable.

If points represent two or more different entities, ensure that the symbols used for each type of entity are clearly distinct, even if the display is reduced. As in line graphs, +, X, open-circle, and solid-triangle symbols are desirable because they remain distinct even at small sizes (Schutz, 1961b). In addition, Chen (1982) presents evidence that symbols that differ in their topological properties (distinguished by a hole, connected components, or a closed form) are more discriminable than symbols that share topological properties (e.g., a solid circle and square). However, the difference between open and solid symbols of the same shape often is difficult to discern, especially after photoreduction.

In addition, Cleveland and McGill (1984b) suggest that point symbols can be ordered on the basis of discriminability as follows: colors, amounts of fill, different shapes, and different letters. However, when Lewandowsky and Spence (1989) asked participants to select two of three point-clouds in a scatterplot and decide which depicted the higher correlation, they found no differences in accuracy when clouds were specified with the different symbol types. Lewandowsky and Spence did find that participants responded more quickly with color than with shapes or fill, and required the most time when confusable letters (H, E, and F) were used as symbols. But if letters were highly discriminable (H, Q, and X), participants could use them faster than circles with different amounts of fill and could use them about as well as different shapes (circle, triangle, square).

Lewandowsky and Spence (1989) also found that participants required the same amounts of time for scatterplots with ten points and with thirty points, which suggests that participants did not read individual symbols in their task. Keep in mind that the specific results often depend on the particular task, choices of colors, shapes, and so forth. Lewandowsky and Spence provide a table of confusions between letters, allowing the selection of highly discriminable letters to use as points (see also Geyer & DeWald, 1973).

Do not indicate overlapping points with different symbols.

Where points overlap, the **do** graph uses larger dots, whereas the **don't** version uses different symbols. Which is easier to read? As illustrated in **do**, a darker or larger dot, or a double dot, better conveys the impression that there is more of something than does the difference between a triangle and a circle. The physical characteristics of the marks should be analogous to the quantitative information, as dictated by the Principle of Compatibility. Recall, however, that the visual system distorts differences in area: Use at most four differences in dot size to indicate different ranges of overlapping points; the sizes should differ by multiples of two to be immediately discriminable. A key or caption stating the values of the different-sized dots is necessary.

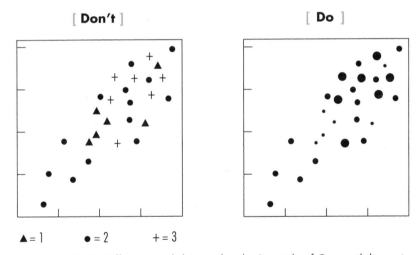

[**Don't**] [**Do**]

▲ = 1   ● = 2   + = 3

Figure 6.12. Different-sized dots exploit the Principle of Compatibility, using "more is more" to indicate overlapping points; the use of different symbols requires the reader to memorize a completely arbitrary convention.

Ensure that error bars do not make less stable points more salient.

If it is important for your message to include error bars, then put a light circle around each point or a light "I" error bar, as in **do**, to indicate a range of variation; the error indication must be light, or less stable points (those with larger variability) will be specified by a larger visual change and hence will be more salient—which is exactly what they should not be, because they are in fact less, not more, reliable.

[**Don't**]   [**Do**]

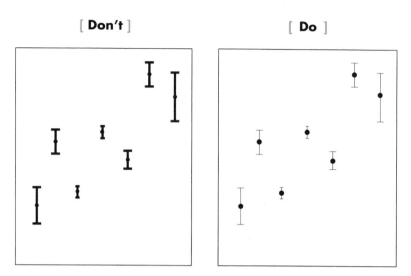

Figure 6.13. Having prominent error bars in fact draws the reader's attention to the least reliable points, those that have the largest bars.

Ensure that best-fitting lines are discriminable and salient.

Best-fitting lines often are supplied in scatterplots to indicate the trend in the data. In the example, the Y axis specifies the number of times a criminal was arrested in 1990, and the X axis the number of times a criminal was arrested in 1991, and each dot represents a convicted criminal. For any good-sized city, you would have a cloud of points. You then might fit a line through the cloud (see appendix 1), which would rise from left to right. If two levels of a parameter (e.g., men and women) are plotted, and best-fitting lines are provided for each, make sure that the lines are easily discriminable (but see pages 60–62 before plotting more than one data set). It is not a good idea to use patterns of dots and dashes to distinguish the lines, however, because the patterns might be confused with the point symbols. Different colors are best, if possible.

If you include a best-fitting line or lines, ensure that the summary statement your graph makes is salient by making the line or lines at least as heavy as the points, as in do. The major purpose of a best-fitting line is to ease the processing load on the reader, and the line will not serve this end if it is not seen quickly and easily.

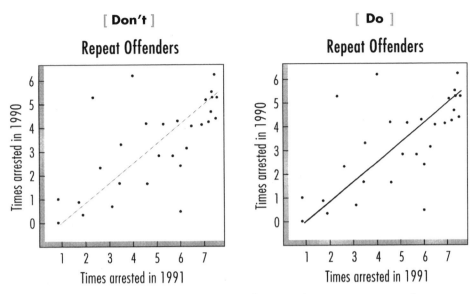

Figure 6.14. A best-fitting line cannot be a good summary statement if it is not immediately visible.

Do not fit a line by eye.

Estimating the best-fitting line by eye is not a good idea. Mosteller, Siegel, Trapido, and Youtz (1985) asked participants to fit lines through scatterplots by eye and found that the average slope of these lines was closer

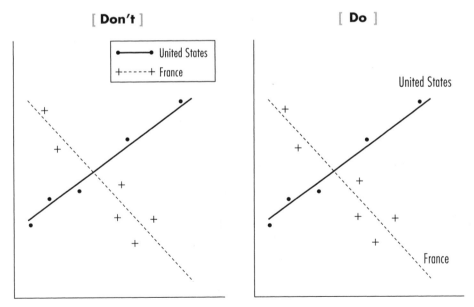

Figure 6.15. If the data permit you to plot more than one set of data, ensure that best-fitting lines are labeled, preferably directly.

to the slope of the "major-axis line" than to the appropriate least-squares line. The major-axis line minimizes the sum of squares of perpendicular distances, not the (correct) distances along the Y axis (see appendix 1 for a discussion of least-squares fits). Similarly, Bailey (1996, cited in Wickens & Hollands, 2000) asked people to fit lines by eye to scatterplots and found that people generally fit the lines at too steep a slope.

If a best-fitting line is included in the display, identify the method of fit in the caption. The reader will not be able to interpret the meaning of the line without this information.

If more than one best-fitting line is present, label each directly.

Include a label in the display for each best-fitting line, as in **do**. Position the label closer to the line it labels than to any other line, using proximity to produce the appropriate grouping.

The Next Step

Turn to chapter 7 for recommendations on shading, hatching, color, three-dimensional effects, inner grids, and other optional features.

Creating Color, Filling, and Optional Components

7

YOU HAVE SELECTED a display type and created the display. However, even if you have followed the recommendations, the components may be difficult to discern, and therefore the graph difficult to read. Hatching, shading, or color can be used to make regions—wedges of a pie, bars, layers—distinctive. Color, as you will see, has further applicability. These techniques not only make a display easier to read, they also enhance its visual appeal. Three-dimensional effects can also make the display more attractive, but unless used carefully, they can muddle the message. Other ways of clarifying your display may be an inner grid, background, caption, or key. Recommendations for creating these features (as well as for multipanel displays, which consist of more than one graph) are given in this chapter.

Color

Color not only adds visual interest to a display but also can help you to communicate effectively. Color may even make a visual display more compelling. Brockmann (1991), for example, cites one case where adding color increased sales nearly four-fold. Why? In this particular case, fully sixty-three percent more people noticed the advertisement when it was in color. Color does grab one's attention (in accordance with the Principle of Salience). If used improperly, however, variations in color can become an active impediment to clear communication.

Color is an enormously complex topic, for at least the following reasons (Christner & Ray, 1961; Hitt, 1961; Travis, 1991). First, color is not a single entity but can be broken down into three aspects:

1. The *hue* (what we usually mean by color) is its qualitative aspect, which depends on the wavelength of the light (from long at the red end of the spectrum to short at the violet end).
2. The *saturation* is the deepness of the color (which can be varied by the amount of white that is added).
3. The *intensity* is the amount of light that is reflected (if shown on a printed page) or that is emitted (if the display is projected from a slide or shown on a computer screen)—in either case, intensity can be varied by the amount of gray that is added.

Second, the perception of hue depends in part on the surrounding colors, and so the display must be looked at as a whole.

Third, people vary greatly in their perception of color. A sizable percentage of the population is color-blind (i.e., they have trouble distinguishing certain colors, usually red and green); also, people's assessments of how intense two colors must be to seem equally bright vary greatly.

The following recommendations will help you to ensure that the colors you use will convey information effectively to all readers.

Use colors that are well separated in the spectrum.

To increase discriminability, colors that are juxtaposed in your display should be ones that are separated by at least one other noticeably distinct color in the spectrum, depicted in the color wheel. Color is registered initially by three types of receptors in the eye: One type is most sensitive to the wavelength seen as orangish-yellow, another is most sensitive to the wavelength seen as green, and the third is most sensitive to the wavelength seen as violet (Reid, 1999; note that these colors are not exactly red, green, and blue, as is commonly believed). We can perceive a gigantic number of colors because, although each type of receptor responds most strongly to one wavelength, all three respond to some extent to any wavelength, but with different degrees of enthusiasm. It is the mixture in outputs from all three kinds of receptors that conveys the precise color. If the mixture produced by two colors is too similar, the colors will not be discriminated.

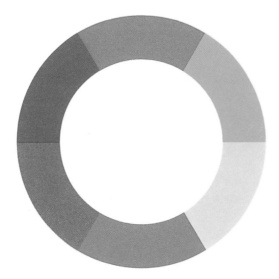

Figure 7.1. The color wheel, which represents our psychological perception that color is not a continuum.

The six colors that people with normal color vision perceive as being most distinct are reddish-purple, blue, yellowish-gray, yellowish-green, red, and bluish-gray. The "eleven colors that are never confused" (with each other, by people with normal color vision) are white, gray, black, red, green, yellow, blue, pink, brown, orange, and purple (see Boynton, Fargo, Olson, & Smallman, 1989; Smallman & Boynton, 1990; see also, for precise specifications of these colors as well as detailed advice about how to produce them, Travis, 1991). Nevertheless, my strong advice is not to use all eleven in a single display: Aesthetics aside, people can generally keep track of only nine colors at the same time (Condover & Kraft, 1954).

Color, if properly used, can help the reader to identify different parts of a display (see, e.g., Aretz & Calhoun, 1982; Casali & Gaylin, 1988) and can help the reader to associate different parts easily (e.g., lines and their labels).

Schutz (1961b) found that participants could better isolate points on specific lines if color was used to distinguish the lines. However, color helped them to find the line with the highest value at a specific point on the X axis only when the lines in a single display could be easily isolated; when the lines crossed many times, the participants located the black-and-white ones as easily as the colored ones. In general, black-and-white displays with highly discriminable lines and symbols were almost as good as color ones; color can often be a useful way of making components of the content distinct but does not always provide an advantage over black-and-white displays (see, e.g., Tullis, 1981).

Make adjacent colors have different brightnesses.

Our visual systems have difficulty registering a boundary that is defined by two colors that are of the same brightness, as in **don't**. Brightness is the psychological impression of intensity, but whereas intensity can be measured by a light meter, our eyes and minds do not accurately translate intensity into brightness (Cavanagh, Anstis, & Mather, 1984). When colors have the same objective intensity, we see blue as the brightest color, followed by red, green, yellow, and white.

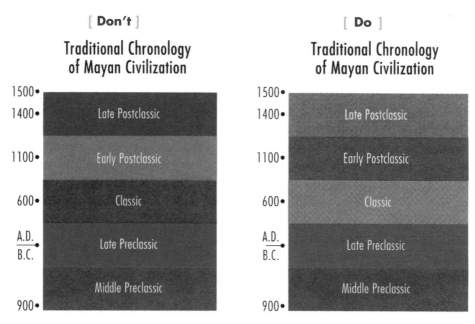

Figure 7.2. Equal brightnesses make boundaries harder to discern. A given level of intensity appears of different brightnesses to different people, and so differences in intensities must be large enough that all viewers will perceive distinct differences in brightness.

Brightness must be adjusted subjectively, and it should vary by quite a bit, for two reasons. First, people vary considerably in their perception of how bright a color is. Second, different sorts of room lighting affect the brightness of different colors in different ways. For example, there is a greater difference in the brightnesses of blue and yellow under fluorescent lights than under incandescent lights (blue seems brighter under fluorescent lamps); similarly, blue seems increasingly brighter than red as the lighting dims.

Travis (1991) points out that the oft-cited admonition not to use white letters on a yellow background (or vice versa) or black letters on a blue background (or vice versa) is in fact incorrect. The problem is that the usual way of producing those colors on a screen results in their having very similar brightnesses; if the brightnesses are varied as much as they are for other pairs of colors, it is evident that the hues are not the culprit. The lesson is to ensure that different brightnesses are used for adjacent colors.

Make the most important content element the most salient.

According to the Principle of Salience, larger differences will be noticed first, so ensure that the most important wedge, object, or segment stands out the most. If no one element is most important, all should be equally salient. After you adjust the brightnesses so that the colors will be easily distinguished, adjust the saturations (the deepness of the color) for the different hues you are using until none predominates. (Note that when a display composed on a computer is printed, the color of the ink usually is not exactly the same as the color on the computer screen.)

Use warm colors to define a foreground.

As discussed in chapter 1, any visual stimulus projects slightly different images into each eye because the eyes are in different places on the head. By the process of stereo vision, the disparity between these images is used by the brain to infer distance. Because each eye is actually aimed slightly askew when we look straight ahead, the lens of the eye acts like a prism. Incoming light is thus broken into the colors of the spectrum, which are projected to slightly different locations on the retina. The psychological effect is similar to what occurs when one object is actually closer than another: Some colors are perceived as "closer" than others. A "warm" color, such as red or orange, will appear to be in front of a cooler one, such as green, violet, or black.

This "false stereo" effect can be advantageous if used in accordance with the Principle of Salience to highlight a feature of the display. If misused, however, the effect can be distracting: A warm-colored object drawn behind a cool-colored one will struggle to move to the foreground, as in **don't**. Moreover, Travis (1991) points out that the depth effect may actually reverse with low illumination (when the pupil is large, and so the eye focuses differently), which means that you must think about the precise context when juxtaposing warm and cool colors. For more on this fascinating phenomenon, see Allen and Rubin (1981), Held (1980, especially p. 86), Travis (1991), and Vos (1960).

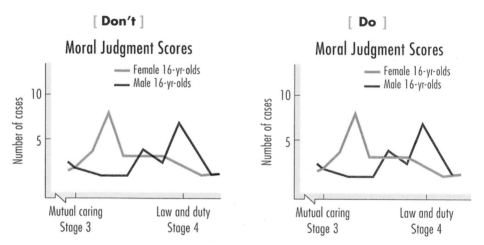

Figure 7.3. A red line that appears to be struggling to move to the foreground produces an effect that is neither esthetically nor functionally desirable.

Avoid using red and blue in adjacent regions.

The lens of the eye, unlike lenses in good cameras, cannot properly focus two very different wavelengths at the same time. This is why red (a relatively long wavelength) and blue (a relatively short wavelength) will seem to shimmer if juxtaposed, as in **don't**. Thus, even though red and blue are easily distinguishable, and thus can be used to make content elements (e.g., lines, bars, or wedges) distinct, don't use them to fill adjacent regions. Also, it is generally a good idea to avoid deep, heavily saturated blue; the eye cannot focus the image properly (it will fall slightly in front of the retina), and so deep blues will appear blurred around the edges. Similarly, avoid cobalt blue, which is in fact a particular mixture of blue and red; for people with normal vision, this color can never be fully in focus because the eye cannot accommodate both frequencies at the same time. The halo you have probably noticed around blue street lights at night is not fog; your eyes are failing to focus the image properly. Finally, it is often a good idea to avoid using red and green to define a boundary because about eight percent of the male population has trouble distinguishing these colors (see Pokorny, Smith, Verriest, & Pinckers, 1979).

[**Don't**]   [**Do**]

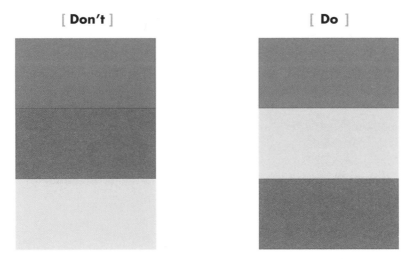

Figure 7.4. The eye cannot properly focus upon red and blue at the same time, so the boundary between these colors shimmers.

Respect compatibility and conventions of color.

If you must use characteristics of color to convey variations in quantity, respect the Principle of Compatibility: Equal increments of quantity should be conveyed by psychologically equal changes in color (hue, saturation, and/or intensity). If increments are not equal, the reader will assume that more similar quantities are conveyed by more similar colors. Given all the foregoing recommendations, respecting this dictum is no mean feat. In addition, objects and events sometimes have predominant colors; do not use color in a way, as in **don't**, that contradicts compatibility.

[**Don't**]   [**Do**]

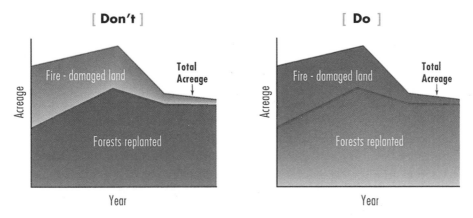

Figure 7.5. The Principle of Compatibility extends to the use of color.

Color can also be used symbolically, but colors can have different meanings in different cultures and even in different subcultures. Brockmann (1991), for example, notes that green means "safe" for process engineers, but green means "infected" for health workers and "profitable" for financial managers. To be on the safe side, test your color choices with people who represent your audience.

For a general sense of the connotations of colors, the best recent survey I have found on this topic was conducted by Hallock (2003), who asked people to associate a color with each member of a list of concepts (e.g., he asked them to indicate which color "best represents trust"; for an earlier survey, see Birren, 1961). Here are some of the salient results just regarding blue and red (colors that the American media has adopted to distinguish between Democratic and Republican states in the United States, referring to states as "blue" or "red," respectively). I first list the concept, then the percentage of people who associated it with blue, and then the percentage of people who associated it with red.

Concept	Associated with Blue %	Associated with Red (%)
Trust	34	6
Security	28	9
Speed	1	76
Cheap/inexpensive	1	9
High quality	20	3
High technology	23	3
Reliability/dependability	43	3
Courage/bravery	28	22
Fear/terror	0	41
Fun	5	16
Favorite color	42	8
Least favorite	0	1

In short, blue generally appears to be a more desirable color, unless fun is at stake or speed is of the essence! However, I stress that these sorts of norms are likely to vary for different groups, times, and places. Indeed, Hallock (2003) illustrates this point simply by breaking down "favorite color" by gender: Whereas thirty-five percent of females prefer blue and nine percent prefer red, fifty-seven percent of males prefer blue and seven percent prefer red. Moreover, the "favorite color" shifted dramatically with age, with more than eighty percent of those older than seventy claiming that blue was their favorite color.

If compatibility (including the aspect of cultural convention) is not relevant, it often can be difficult to decide which color should stand for which entity. Travis (1991, pp. 123–124) formulates six principles of color coding that can guide you in such situations. These principles mesh nicely with the principles developed in this book (which are noted in parentheses following each of his principles): Colors must be discriminable (Discriminability); colors must be detectable (Discriminability); colors must vary in psychologically equal steps—if two colors are relatively similar, readers will believe that the things they represent are related in some way (Compatibility and Informative Changes); colors should be meaningful—do not vary them unless they reader will know what the variations mean (Informative Changes and Appropriate Knowledge); and colors should be aesthetically pleasing (use as few as possible and ensure that they do not vibrate or clash).

Travis also suggests using no more than five different colors in a display; this recommendation also follows from the Principle of Capacity Limitations (specifically, those aspects concerned with limited short-term memory capacity; I do not offer it as a separate recommendation here because it is subsumed by my recommendations pertaining to the amount of content that should be displayed in a single graph). Trumbo (1981) relies on some similar ideas to suggest ways to use color in maps to display more than one type of information at a time.

Use color to group elements.

The grouping law of similarity implies that regions of the same color will be seen as a group. This principle, used in **do**, is very useful if the reader is to compare two or more elements in different places. If two pie graphs are used, using the same color for corresponding wedges will group them effectively. In addition, Parkin (1983; discussed in Pinker, 1990, pp. 114–115) found that color effectively groups labels with lines, even in complex displays. Wickens and Andre (1990) provide evidence for the importance of color in helping the reader to integrate different types of information, but this evidence is less compelling for bar graphs than for exotic "object" displays (which make use of integral dimensions to pack much related information into a single display). Nevertheless, good black-and-white patterns can be almost as good as color distinctions for grouping and identifying individual elements.

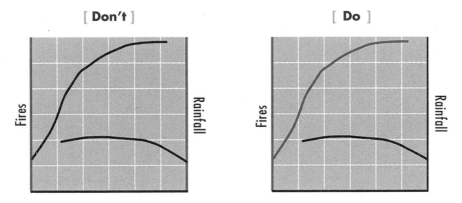

Figure 7.6. If used properly, color can be a very effective grouping device.

Avoid using blue if the display is to be photocopied.

Many of the most popular black-and-white copying technologies reproduce the color blue poorly or not at all. If blue is used, test to be sure that it will reproduce well.

Avoid using hue to represent quantitative information.

Respecting the Principle of Compatibility, do not use hue to represent differences in amounts; use it as a label to promote discrimination and grouping. Look at the **don't** figure: Shifting from red to violet (the spectrum ranges from red, orange, yellow, green, blue, indigo, to violet) does not convey the impression that an amount has been added the way that shifting from a short bar to a tall bar does. In fact, the shift from red to violet does involve more of something (more oscillations in the amplitude of a light wave). Nevertheless, although the wavelengths of the hues order them along the spectrum, the hues themselves are not arranged into a psychological continuum.

For evidence that people have difficulty using differences in hue to extract quantitative information, see Cuff (1973) and Wainer and Francolini (1980). If practical considerations require that you use hue in this way, Travis (1991) suggests using colors that are relatively similar to construct a continuum, such as green, green-yellow, yellow, yellow-orange, orange-red, and red. This is good advice unless you want the reader to detect adjacent val-

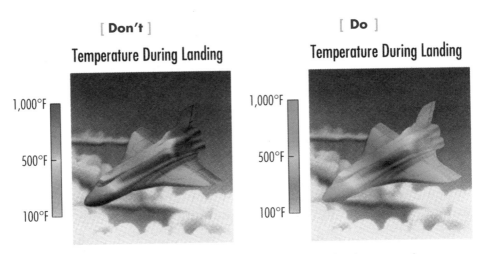

Figure 7.7. Because differences in hue are not immediately perceived as differences in amount, the reader is required to memorize a key if hue alone is used to represent quantities (as in the version on the left). However, if the saturation varies along with hue, as in the version on the right, the reader can easily see differences in represented amounts.

ues at a glance; these colors are sufficiently similar that such discrimination would require effort.

However, if color is used simply to label quantities, hue can convey quantities adequately. Hastie, Hammerlie, Kerwin, Croner, and Herrmann (1996) studied how well color can label quantitative information in maps. They compared four ways of conveying quantitative information within regions of maps, namely, variations in gray scale (from light to dark), variations in brightness levels of a single hue (blue), variations in hues that were ordered from red through violet (in the order of the spectrum), and variations in hues that were randomly assigned to quantities. When participants were asked to extract specific values from these coding schemes, they performed better when given displays that varied hues than when given displays that varied the gray scale or brightness of a single hue. When asked to estimate the mean value across a region, the participants performed comparably with the different sorts of displays (although there may have been a slight advantage for the gray scale and intensity-varied displays).

These results suggest that a specific color is more easily associated with absolute values than is a specific shade of gray. This study did not require ordering the values, however, which is where I would expect hue (which does not "automatically" convey variations quantity along a scale, as noted above) to be inferior to shades of gray or levels of brightness.

Use deeper saturations and greater intensities for hues that indicate greater amounts.

If practical considerations force you to use hue to convey quantitative information, then use deeper saturations (more color) and greater intensities (more light) for hues that indicate greater amounts, as in the **do** illustration of the space shuttle. Based on research results, Cuff (1973) advises using these variables alone to represent quantities, holding the hue constant. We see increases in both of these visual dimensions as increases in amount, and so these increases can signal increasing quantity effectively.

Do not use hue, saturation, and intensity to specify different measurements.

The previous recommendation will also lead you to avoid using hue (the qualitative aspect of color), saturation (the deepness), and intensity (the amount of light) to convey different types of information. For example, do not be tempted to specify three independent variables in a single pie graph (the **don't** version of the ice cream consumption example attempts to use intensity to indicate another variable, the average temperature). The reader will have a very difficult time sorting out the different variables because hue and intensity, and hue and saturation are integral perceptual dimensions. A reader looking at a colored wedge cannot help but pay attention to its deepness and intensity as well as its hue. Like height and width, these properties should not be varied independently to convey information about two variables.

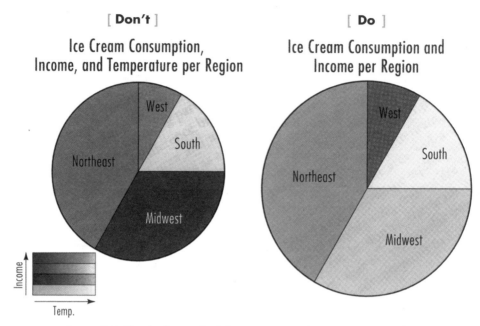

Figure 7.8. The display on the left uses size to indicate consumption by region, hue to indicate income, and saturation to indicate temperature; it is a puzzle to be solved. The display on the right uses only size and saturation to convey amounts, and readers can clearly sense the orderings here.

Hatching and Shading

Individual wedges of pie graphs, elements of visual tables, or segments of divided bar graphs are often filled with hatching or shading. We must take care to consider the spacing between texture elements, to ensure both that different regions can be readily distinguished and that the patterns are not visually irritating. You might wonder which is better, hatching or shading. To see why this is a difficult question to answer, consider the following: We want to know whether children find shape or color more important, and so we vary the shapes and colors of forms and ask the children to sort them into two piles. We could induce the children to use shape if the forms are circles versus triangles and the colors are subtly different shades of red, and could induce them to use color if the shapes are slightly different ellipses and the colors are blue and yellow. The problem is how to equate the particular values on different dimensions. This is not as easy as it might sound because context affects the way we perceive specific stimulus values; for example, the way we see a given color depends in part on the other colors that surround it (see Travis, 1991). Thus, we cannot assert in general that one sort of variation (such as in hatching or shading) is better than another.

Ensure that visual properties are discriminable.

The Principle of Discriminability states that two or more marks must differ by a minimal proportion in order to be distinguished from one another; the critical factor is the proportion, not the absolute amount. When you use intensity (which we see as brightness) or texture (e.g., cross-hatching) to define different wedges or segments, remember that the visual system does not simply record what it sees point for point, like a television camera. Rather, it tries to delineate edges and boundaries, both between objects and their backgrounds and between individual parts of objects. The brain does this by responding to differences in contrast, color, texture, orientation, and other properties between neighboring regions. Whenever the brain registers a significant difference between, for example, a light area and a dark area or a striped area and a polka-dotted area, it does the equivalent of drawing a line that serves as a boundary. If neighboring regions are too close in lightness, texture, or color (as they are in **don't**), the brain will not establish a border between them—that is why a polar bear lying on the snow is hard to see. Camouflage is advantageous to a polar bear but not to a display.

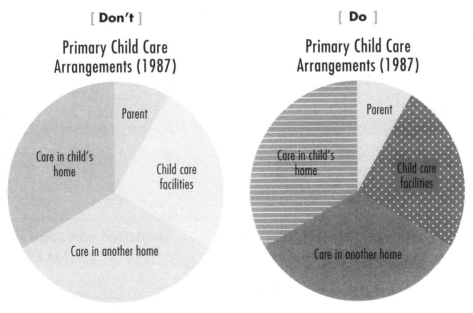

Figure 7.9. Poor discriminability of regions, confusing enough here, is particularly damaging if the reader is to compare corresponding wedges of a multipanel display.

Ensure that orientation varies by at least thirty degrees of arc.

When you read an analog clock (one with hands, not just numbers), some positions of the hands require closer attention than others. It may be no accident that clock faces are divided into 12 equal and easily discriminable increments. The Principle of Perceptual Organization (specifically, the aspect of input channels concerned with orientation sensitivity) states that if differently oriented hatchings are used to distinguish regions, they should be at least thirty degrees apart (which is the angle formed by the hands of a clock when they point to adjacent numbers, as at 12:05). Various studies of discrimination of lines at different orientations have shown that when lines differ by at least thirty degrees, as in **do**, we can distinguish among them without having to pay close attention.

The estimates of the specific degree of orientation tuning of the "channels" are somewhat variable, depending on the precise technique used to measure orientation and the testing conditions (see, e.g., De Valois & De Valois, 1988, ch. 9). The estimate I offer here is somewhat conservative (some researchers suggest that individual channels respond to orientations within a range of about twenty degrees) but is consistent with relevant findings of Nothdurft (1991). Nothdurft examined the so-called pop-out phenomenon, which occurs when stimuli are sufficiently different from the background that one notices them immediately. He found that lines will seem "pop out" from a field of other lines if their orientations differ by about thirty degrees. Neurons at various levels of the visual system are tuned to respond only to

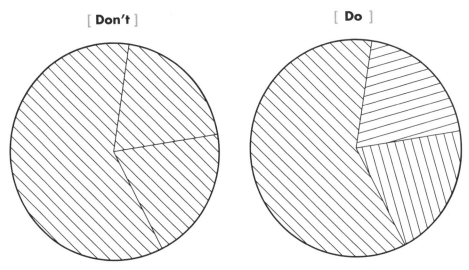

Figure 7.10. Line orientations must be immediately discriminable to delineate regions clearly.

edges or shapes at particular orientations; in the initial stages of processing, the neurons have rather broad "tuning curves" and so respond to edges within a relatively wide range of orientations (the mind is not a camera). Thirty degrees is large enough to ensure that different input neurons will be used to encode patterns of different orientations.

That said, although discriminable, the **do** stripes are hard to look at. Avoid such jarring displays; color or shading is usually better than such patterns of stripes.

Vary the spacing of texture patterns with similar orientations by at least a ratio of 2 to 1.

The example illustrates the salubrious effects of alcohol (by some measures) on Londoners. It is immediately obvious from **do** that there are more nondrinkers in poor health than in good health, and vice versa for moderate—and even heavy—drinkers. The **don't** version is not so easy to read. If cross-hatching, stripes, dots, dashes, or other regular patterns have similar orientations (i.e., within thirty degrees of each other), the densities of the pattern should differ by at least 2 to 1. If a region has eight hatch lines to the inch, to be immediately discriminable adjacent regions should have either four or fewer lines to the inch, or sixteen or more lines to the inch.

This recommendation is based on the discovery that the visual system operates at multiple levels of acuity (see pp. 8–10), as if we had different lenses. What we see comes to us from a number of "input channels," which register regions of different sizes (and thus pick up different amounts of detail—less detail is detected when larger regions are monitored). Furthermore, there are several types of each channel, tuned for different orientations. All these channels operate at once.

The acuity of a channel is described by its *spatial frequency*, the number of regular light/dark changes (e.g., light and dark stripes of equal width) that fit into a specific amount of the visual field. Spatial frequency is measured in cycles per degree of visual angle. Two degrees of visual angle is roughly equivalent to the apparent width of your thumb when you hold it up at arm's length and look at it with one eye. A cycle is a complete sequence of the light/dark change. A spatial frequency of two cycles per degree, then, would have two dark–light pairs for every extent that is half as wide as your thumb appears when you hold it up at arm's length. De Valois and De Valois (1988, ch. 6) note that a channel will barely respond (if at all) to a variation in spatial frequency of 2 to 1 above or below its preferred spatial frequency.

Thus, when we look at an area of evenly spaced hatch lines, dots, or patches, we will also mentally take in, like it or not, areas of other similarly oriented lines, dots, or patches whose cycles range from half as frequent to twice as frequent. The closer the spacing is to that of the area we are paying attention to, the harder it is to ignore. Therefore, to be immediately discriminable, two repetitive patterns (e.g., stripes or dashes) that have similar orientations should be registered by different channels, which means they should differ by more than 2 to 1.

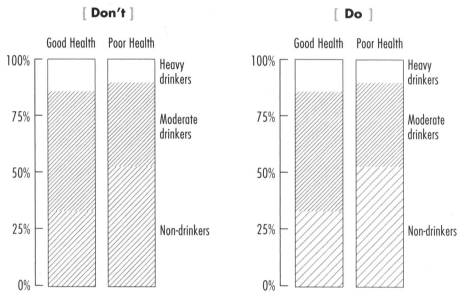

Figure 7.11. If segments are not immediately discriminable, the reader has to work harder than necessary to compare corresponding elements.

Avoid visual beats.

Similarly oriented patterns that are registered by the same channel—those whose spatial frequencies do not differ by at least 2 to 1—may appear to shimmer, a distracting and irritating effect. This shimmering is analogous to what happens in music if two notes are only slightly different, so that their harmonics will periodically reinforce each other, causing pulsating beats. This is a helpful cue if one is trying to tune a guitar but an unwelcome irritation if one is listening to a song being played. Visual shimmer, illustrated here, results when brain cells try to organize input into homogeneous regions, an attempt that is not clearly successful or unsuccessful if the patterns are very similar but not identical. There is no simple way always to predict when juxtaposed hatch marks will appear to oscillate, creating an op-art pattern. If you ensure that the spatial frequency spread respects the Principle of Perceptual Organization, the specific aspect of levels of acuity, the occurrence of this phenomenon is less likely.

Figure 7.12. An annoying shimmer occurs when your visual system is struggling to detect a poorly defined edge.

Three-Dimensional Effects

When we look around us, the cues our visual systems use to assign depth allow us to see the world as it is—in three dimensions. When the visual system, relying on the same cues, addresses a two-dimensional surface, it can be fooled into seeing a third dimension that is not present (Kaufman, 1974). If at all possible, we will interpret a pattern as representing a three-dimensional object. This fact has been exploited to produce pseudo–three-dimensional displays, which seem to have become increasingly popular (perhaps in part because they are easy to produce with widely available computer programs). (Because "pseudo–three-dimensional displays" is such a mouthful, I'll abbreviate by saying simply "three-dimensional"—but keep in mind that these displays are not actually three dimensional: They all are two dimensional, drawn on paper or presented on a screen.) Three-dimensional displays can be useful if they bring the reader's attention to the important comparisons being made, but they can be harmful if they obscure the message. You must decide whether the added visual interest is worth the risk that information will be lost.

Any of the graph formats we have discussed can be drawn to appear three-dimensional, with depth indicated along an unseen Z axis. It is now common to see bar graphs in which the bars look like square pillars sitting in a partially opened box that serves as the framework. A sheet of paper, or a screen, is two-dimensional, and so special tricks must be used to fool the eye and mind into seeing a three-dimensional object.

Three types of tricks are often used in visual displays, all of them based on environmental cues that usually signal information about depth:

1. *Linear perspective:* The visual system interprets converging lines on a two-dimensional surface as if they were parallel lines extending in depth.
2. *Texture gradient:* In a photograph of a field of wheat or a brick road receding into the distance, the texture—the density of the stalks or bricks—increases in the portions of the scene that are in reality (although not in the photograph) farther away. Our visual systems assume that as the texture elements become more tightly packed together, that part of the scene is more distant.
3. *Occlusion:* Closer objects can obstruct our view of farther objects, but not vice versa—and so partial occlusion (i.e., obstruction) is one piece of information our visual systems use to determine relative distance. If a completely closed shape interrupts the boundary of a second, then the first shape is seen as standing in front of the second, creating a sense of depth.

You can often exploit these cues to convey depth effectively on two-dimensional pieces of paper or computer screens. The Principle of Perceptual Organization (specifically, the aspect concerned with three-dimensional projection) states that patterns will be seen as three-dimensional whenever possible. Moreover, the visual system operates as if it assumes that the same-sized image (size measured in degrees of visual angle) produced by something farther away signals a larger object. Thus, in Figure 7.13 the three-dimensional bar (box) at the bottom right looks larger than it would if it were not partially occluded, because one unconsciously compensates for perspective effects.

These three-dimensional cues on two-dimensional surfaces are imperfect substitutes for the real thing, and confusion may arise. The famous drawings of M. C. Escher show how we can be tripped up. Although three-dimensional graphs are rarely as confusing as Escher drawings, they sometimes lean in that direction.

Do not use three-dimensional perspective to communicate precise information.

Although three-dimensional graphs can convey general impressions of trends, they are not very good for representing precise amounts (Casali & Gaylin, 1988). Because what are in fact two-dimensional displays do not exploit all of the depth cues we use in everyday perception (e.g., relative

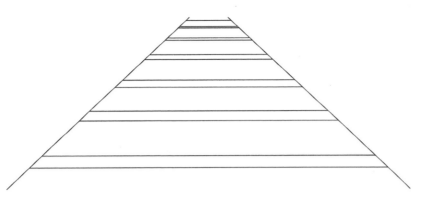

Figure 7.13. The visual system interprets converging lines as parallel lines extending in depth.

The visual system interprets progressively smaller and more tightly packed elements as constant-sized elements extending in depth.

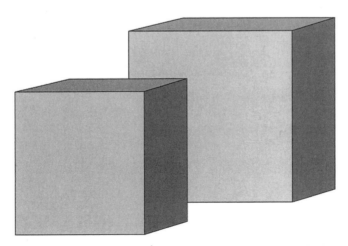

The visual system interprets an enclosed shape that interrupts the boundary of another shape as being in front.

motion, stereo vision, shading), they do not depict three-dimensional information very accurately. In addition, the visual system does not estimate volume very accurately (Stevens, 1974, 1975), so if your display is supposed to convey precise information, keep it two-dimensional.

Several researchers have found that adding the third dimension to a graph actually impairs readers' performance (e.g., Fischer, 2000; Hicks, O'Malley, Nichols, & Anderson, 2003; Siegrist, 1996). This is particularly a problem when complex shapes are used to represent the data, such as the length of spikes on a circular helix (Hicks et al., 2003) or a "wire frame" surface that depicts measurements found with different combinations of values on the different independent variables (Shah, 2002).

However, in spite of such findings, three-dimensional graphs may have their place. For example, Spence (1990) found that participants were equally accurate when comprehending the quantities conveyed in two- versus three-dimensional versions of seven different types of graphs. But the results of studies on this topic are rarely so clear-cut (e.g., Risden, Czerwinski, Munzner, & Cook, 2000). For example, Carswell, Frankenberger, and Bernhard (1991) asked participants to estimate the relative size of a wedge of a pie compared to the whole (for pie graphs) or the relative height of a bar or point compared to a reference bar or point (for bar and line graphs, respectively). They found that these estimates were comparable for two- and three-dimensional versions of pie and bar graphs, but participants were much less accurate with three-dimensional line graphs than they were with the two-dimensional versions. Furthermore, participants in Carswell et al.'s study actually classified trends in three-dimensional line and bar graphs faster than they did in the corresponding two-dimensional versions—but they made more errors. Because the participants responded more quickly when they made more errors, these results are difficult to interpret; they may only indicate that the participants were "jumping the gun," responding before they were ready (but even this interpretation of the results should make us take pause when considering whether to use such displays).

Carswell et al. (1991) also found that although three-dimensional line graphs tend to be recalled more poorly than are two-dimensional line graphs, there was no difference for bar graphs, stacked-bar graphs, or pie graphs (and, in fact, there was a slight trend for three-dimensional versions of pie graphs to be recalled better than two-dimensional versions).

I should note, however, that the precise nature of the three-dimensional display probably affects performance. For example, Siegrist (1996) found that adding the third dimension to bar and pie graphs impaired performance, even though his method and stimuli appear similar to those of Carswell et al. (1991). Siegrist suggests that Carswell et al. used a relatively high perspective angle, which facilitated interpreting the displayed values. Clouding the waters, I must note that Carswell et al.'s displays had fewer bars or segments than those used by Siegrist. Moreover, even Siegrist did not find that two-dimensional bar graphs were always better than the three-dimensional versions. Rather, the impact of the dimensionality of the display depended on the position of the queried bar: In his three-bar displays, the participants tended to underestimate the second bar in graphs that included the third dimension, and actually overestimated the second bar in graphs that were purely two-dimensional.

Similarly, Fischer (2000) found that adding the third dimension to the bars in a bar graph did not affect the time people required to compare them, but adding the third dimension to the framework did slow down the participants. Adding the third dimension to the framework was especially a problem when the bars were also drawn in three dimensions, probably because the participants had difficulty aligning the bars with the corresponding part of the frame (reflecting our imprecision with spatial relations). In addition, Fischer found that participants did not look at three-dimensional bar graphs any longer than they looked at two-dimensional ones, nor did the time to compare bars differ when the participants made their judgments after the display was removed (and hence relied on memory for the bars).

Zacks, Levy, Tversky, and Schiano (1998) examined the effect of two- versus three-dimensional bars when participants picked out which bar in a set had the same height as a sample. Although they found that participants were less accurate with the three-dimensional versions, the effect was very small (and was dwarfed by the effects of the height of the bar per se, with participants making more errors for taller bars). Similar findings were obtained in another experiment, in which participants estimated relative magnitudes instead of matching bars; again, the advantage of the two-dimensional displays was very small. Moreover, the slight advantage for the two-dimensional versions was completely eliminated when participants made their judgments from memory. Although these findings are often cited by other authors as evidence for avoiding three-dimensional renderings, Zacks et al. actually say, "From a practical point of view, it [i.e., the small effect of dimensionality] also suggests that we should pause before making strict design recommendations based on the cognitive-visual problems with 3D graphs" (p. 126; bracketed material added for clarification).

Putting aside for now questions of the actual efficacy of two- versus three-dimensional displays, it may be worth considering the fact that Carswell (1991) and Carswell et al. (1991) found that participants developed more positive attitudes about data when they were displayed in a three-dimensional graph than when they were displayed in a two-dimensional graph. However, I now must hasten to add that Fisher, Dempsey, and Marousky (1997) did not find that their participants preferred three-dimensional graphs over two-dimensional graphs.

Why so much inconsistency? The precise instructions given to participants in these studies may be crucial. Consider a study reported by Levy,

Zacks, Tversky, and Schiano (1996), who asked participants to choose a graph for specific purposes. In one case, the participants were told that the graph was to be used by readers to "understand what is going on" (p. 46). In another case, the graph was to be used to communicate to others. Although the participants overwhelmingly chose two-dimensional graphs in the first case, many more of them chose three-dimensional graphs in the second case.

However, Levy et al.'s (1996) findings are ambiguous. Tractinsky and Meyer (1999) argue that when told that the graph was to be used to communicate to others, the participants responded not on the basis of what sort of display they thought would communicate most effectively, but rather on the basis of what they thought would most *impress* the reader. According to Tractinsky and Meyer, three-dimensional graphs are used as a technique for promoting oneself:

> For example, the use of gratuitous graphics [including those with an added third dimension] may allow presenters to demonstrate their mastery of state-of-the-art technology, or their "profession-alism"; they may demonstrate the extra effort they have put into setting up the presentation or show their care about the content-ment of their audience. Thus, social considerations in organizations are likely to promote self-presentation behavior in the context of information presentation. (p. 401; bracketed material added for clarification)

To test their interpretation, Tractinsky and Meyer (1999) asked partici-pants to choose a display that would be best suited for a specific purpose. They found that the participants preferred two-dimensional graphs as an aid to making decisions, both when one only was making the decision oneself *and* when the graph was to be shown to others who were making the same deci-sion. In contrast, when the point was to impress others, participants preferred graphs that were drawn to include three-dimensional perspective.

In an interesting twist, in one of their studies Tractinsky and Meyer (1999) also found that participants preferred three-dimensional graphs when the data were unfavorable. The researchers interpret this preference as a kind of compensatory behavior, in which the presenter is trying to show that he or she is competent even in the face of bad data. However, Tractinsky and Meyer also concede that the "gratuitous graphics" (specifically, the added third dimension) could be used to obscure the actual, unfavorable data. And, of course, both motivations may underlie one's preferences for "fancy graph-ics" when the data are less than ideal.

The bottom line from this research, in my view, is that you should stick to two-dimensional displays if you want to convey precise amounts, par-ticularly if it is likely that the reader will see the display only for a brief pe-riod of time (as may occur in a PowerPoint presentation). However, if you can give the reader ample time to inspect the display, and you mostly care about the reader's general impression, emotional reaction, or subsequent memory for the display, then three-dimensional pie and bar-graph variants are acceptable. However, I have yet to see evidence that three-dimensional line graphs should ever be used instead of the corresponding two-dimensional versions.

Use views that allow the reader to see the entire content.

Occlusion, one of the depth cues, can result in loss of information—as occurs in both of the illustrations here (these both are **don't** displays). Make sure that all critical aspects of the content (e.g., the tops of bars) are visible. Avoid occlusion if it distorts the visual impression of the size of a region. It often is best to depict the display as if it were seen from the front, from a position elevated just enough to see over the tops of the "nearer" elements. Avoid adopting a point of view that is too high; top views produce foreshortening, which obscures the relative heights of content elements. And if the point of view is too low, the closer elements obscure the farther ones. It often is useful to expand the height of the graph (which is the same as contracting the X and Z axes); the graph will appear more three-dimensional, and differences in height will be easier to discriminate.

Figure 7.14. Three-dimensional surfaces should not be depicted from an angle so high that depth is lost (left) or from an angle so low that closer surfaces obscure farther ones (right).

Show all parts of the display from the same viewpoint.

Can you determine from the **don't** graph the average salary of major-league baseball players in 1985? One of the reasons this is difficult is that the content and background are seen from different points of view. Depict all parts of a three-dimensional display from the same viewpoint; the visual system assumes that an object is seen from only a single vantage point, and considerable effort will be required to reorganize the display mentally if it is not. Similarly, never have one part of the display drawn in two dimensions with a three-dimensional part tacked on. If these recommendations are violated, it will be very difficult to read a value off the display.

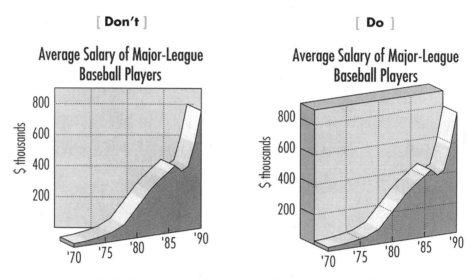

Figure 7.15. Values can be nearly impossible to read when all components of the display are not viewed from a single perspective.

Avoid see-through displays.

See-through displays such as **don't** show what we would see if we had X-ray vision; nearer surfaces are treated as if they were transparent, allowing farther surfaces to be seen through the nearer ones. These displays are rarely effective because the reader organizes the parts incorrectly; the grouping laws result in spurious groupings. If you must use such a display, use a different color, shading, or line thickness for the underside of a surface if any part of it is visible. Better yet, make the walls of a surface opaque, as in **do**, by shading or by coloring them in a solid color.

Huber (1987) describes an exception to this recommendation. He used moving displays on a computer screen to present multivariate scatterplots and reports that viewers can easily find outliers and clusters in such displays, as well as locate the most informative viewpoint (see also R. G. Miller, 1985; Reaven & Miller, 1979). However, he also reports, "In general, it was our experience that only simple minded approaches will be used, and be interpretable, by anyone other than the inventor of the method" (Huber, 1987, p. 451).

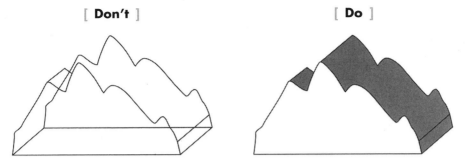

Figure 7.16. See-through displays often lead readers to group components incorrectly and are generally difficult to interpret.

Use a vertical wall for the Y axis.

It is almost impossible to read off the number of major corporate head-quarters from **don't**, whereas it is possible (with work!) to do so from **do**. Use a vertical wall for the Y axis, which will appear as a plane projecting in depth along the Z axis, instead of a single line. But even in this case, the reader will be able to read specific Y values only with considerable effort; the three-dimensional cues are not good enough to allow the reader to locate the appropriate position in depth easily. If the reader is supposed to extract precise values, label these values with numbers. But if you really want to help the reader find the information with less work, use a two-dimensional display instead.

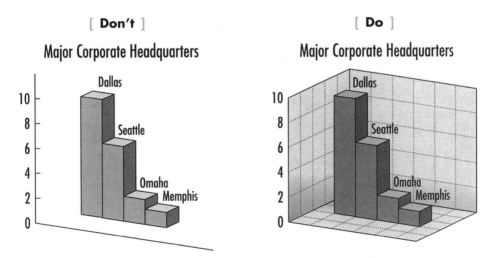

Figure 7.17. A vertical wall allows readers to extract more precise values than would be possible without it.

If the content is a three-dimensional sheet, cover it with regularly spaced grid lines.

Some three-dimensional graphs include a different independent variable along the X axis (running perpendicular to the line of sight, as usual) and along the Z axis (running in depth). The content is a sheet, which typically appears as a landscape with peaks and valleys. Because the local convergence of lines is a good depth cue, the three-dimensional effect will be enhanced if the sheet looks like a surface that, as in **do**, is covered with a relatively dense grid pattern.

[**Don't**]   [**Do**]

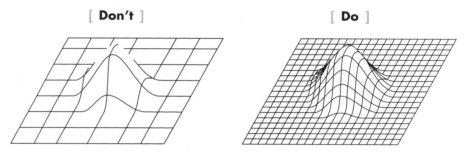

Figure 7.18. Grid lines can create a three-dimensional surface, but the effect falters if the lines are not packed closely enough.

Inner Grid Lines

An inner grid can be used with any format that includes a framework. Vertical grid lines help the reader to isolate a specific place on a line, and horizontal lines help the reader to spot particular values of the dependent measure. If the inclusion of such lines seems appropriate (as determined by the guidelines discussed in chapter 2), create them according to the following recommendations.

Make inner grid lines relatively thin and light.

Which version of "Champagne Sales" would you rather use to find out how much bubbly was sold in 1986? The inner grid lines should be discriminable from those that outline the framework or content elements. They should not be so salient that they distract, nor should they obscure the content material.

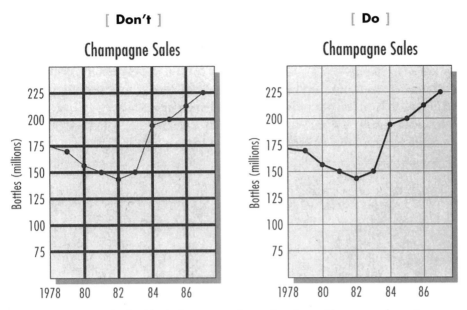

Figure 7.19. Grid lines that are too heavy interfere with content elements.

Use more tightly spaced grid lines when greater precision is required.

The Principle of Relevance implies that grid lines should specify only the level of precision necessary to answer the appropriate questions. As a rule of

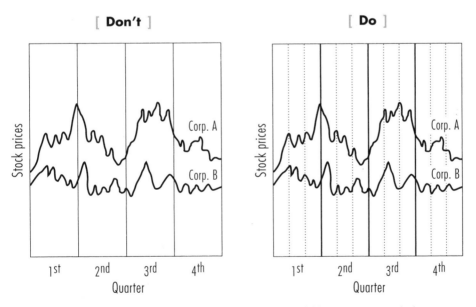

Figure 7.20. The frequency of the grid lines should be consistent with the fluctuations of the curves; if the point is to present quarterly, not weekly, prices, the data should have been averaged and a single number plotted for each corporation for each quarter.

thumb, consider locating the lines at increments that are about four times the acceptable rounding error (i.e., level of precision), as in **do**. If you want the reader to be able to extract figures to the nearest week, grid lines placed at monthly increments, as in **do**, should be sufficient. We can estimate length quite well and can visually divide a line into four roughly equal segments. (Carter, 1947, reports one of the few studies of the effectiveness of grid lines. Unfortunately, the previous recommendation—"Inner grid lines should be relatively thin and light"—was not respected, and the graph with denser grid lines also lowered the contrast of the content line. Thus, we cannot take at face value the finding that very frequent grid lines did not increase accuracy.)

Insert heavier grid lines at equal intervals.

Which version of "Percent of Children in Poverty" is easier to use to determine the years in which twenty percent of children were under the poverty line? If tick marks vary in size or darkness, you should line up a relatively heavy grid line with each darker tick mark. If grid lines are closer than one centimeter apart, it is a good idea to make some of them heavier (Beeby & Taylor, 1973). Given our convention of using Arabic numerals, which are based on powers of ten, every tenth grid line should be slightly thicker than the others. If greater precision is required, use an intermediate level of thickness as appropriate. Keeping in mind the Principles of Salience and Discriminability, take care that these lines do not obscure the content. Varying line weight this way will help the reader to distinguish a single line, which can then be used to track from the content material to the Y axis.

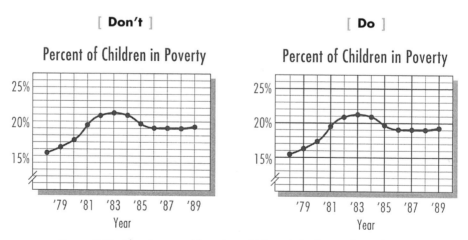

Figure 7.21. The staggered heavier grid lines help the reader to locate specific values.

Inner grid lines should pass behind the lines or bars.

Grid lines should not obscure the content material. The grouping law of good continuation ensures that each grid line will appear to progress across the entire display even though it is physically interrupted by the content material, as in **do**.

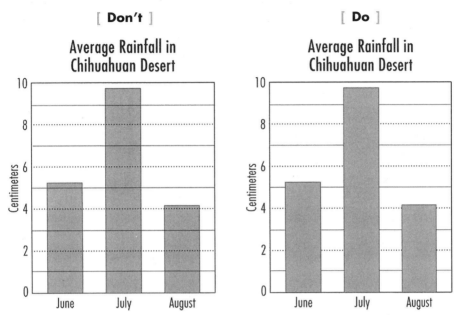

Figure 7.22. Superimposed grid lines interfere with the content and relegate it to the background; in contrast, we see grid lines interrupted by bars as behind them, and the lines push the bars to the foreground.

Background Elements

The background is not an essential part of a display. However, although it is included primarily to make the display more attractive or interesting, it can also be used to label the display or to reinforce the message.

Use a background to reinforce the main point of the display.

When a display is used to catch the reader's attention or to enhance a particularly dramatic point, an effective background underlines the message of the display, providing a kind of pictorial label. The background design must be compatible with the content of the display, as in **do**. Show a couple of people the background image you are considering using and ask them to name it with the first word that comes to mind; this word should be appropriate for the message conveyed by the display, both in its direct meaning and in its indirect connotations (in accordance with the Principle of Relevance). Do not use decorations that convey no information about the topic of the display; according to the Principle of Informative Changes, every element of a display should convey information. Tacked-on elements will only serve to distract the reader.

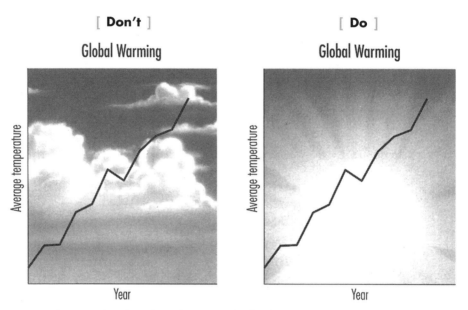

Figure 7.23. The editorial content of a background should not conflict with the message of the display.

Ensure that the background is not salient.

Can you easily tell from **don't** what houses cost in New Hampshire in 1990? Do you initially have trouble distinguishing parts of the drawing of the house from parts of the content itself? The background should not interfere with the information-bearing lines and regions, nor should its salience lure the eye away from these elements. Give the background low contrast and low saturation, few details, and soft edges.

Figure 7.24. If salience leads readers to see the house before the data, they will have to work to sort out the content.

Ensure that background elements do not group with content elements.

Look at the **don't** version: At first glance, it seems that South Korea and India have similar levels of defense spending. But this is not the case, you can see from **do**. Make sure that background figures do not group with

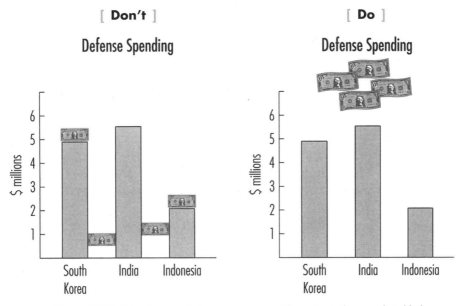

Figure 7.25. If background elements group with content, the graph is likely to be misinterpreted by a casual reader.

parts of the display itself. You should be aware of similarities, proximities, good continuations, common fate, and good-form effects (which are aspects of the Principle of Perceptual Organization, discussed in chapter 1) that can produce confusion between background figures and content.

Captions

A caption is not necessary in all contexts. It often is used when graphs are presented in texts, but rarely when graphs are used during presentations. When a caption is not included, however, its functions—clarifying and directing attention—must be fulfilled by other material that accompanies the display.

Put the caption under the display.

Convention dictates that captions appear under displays. This recommendation can be ignored, however, if you have many displays and adopt a consistent alternative format (e.g., placing the caption under the title or in a margin). But if you decide to use an unconventional design, keep in mind that readers will require more time to read the display until they master your scheme.

Make the caption distinct from the text.

The caption should be easily distinguished from the surrounding text. It should be physically closer to the display than to the text, so that it will group appropriately by proximity. It is also helpful if the caption is typographically distinct from the text, perhaps in a smaller—but still easily legible—size or in a different typeface.

The Key

Use a key to prevent clutter (see the discussion on pages 69–70). The key has two components: *patches*, which are samples of the texture, color, or other visual properties that correspond to the individual components of the content, and *labels* that identify the patches. The labels should be created in accordance with the recommendations given in chapter 3.

Place the key at the top right of a single panel or centered over multiple panels.

In a single-panel display, the key is by convention positioned at the top right; in a multipanel display, it is centered at the top, directly beneath the title. Readers familiar with graphs will expect to find the key in one of these locations. If aesthetic or other considerations lead you to put the key in an unconventional location, realize that the readers may have to search for it, which will tax their limited processing capacity. However, an unconventional location may be acceptable if space is available only in another part of the display, and the key is clearly isolated from the content. In addition, if you consistently place the key in the same location in a number of displays, the reader will soon learn where to find it.

Ensure that labels and patches are detectable and that corresponding labels and patches form perceptual groups.

Ensure that the label and the patch are detectable even after any photo-reduction planned. The principles that ensure that lines and regions are discriminable should be respected (pp. 158–161 and 169–172). Associate the label and the patch by making sure that they are closer to each other than to any other part of the key; the grouping law of proximity will then group the elements properly.

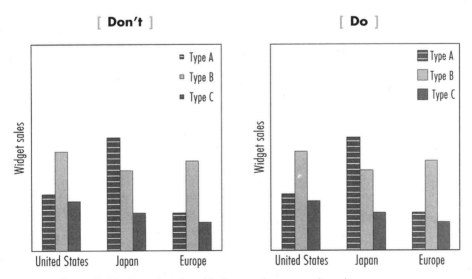

Figure 7.26. An indecipherable key renders a graph useless.

Use the same order for patches and their corresponding content elements.

The patches in the key should be presented in the same order as the corresponding content elements in the graph itself (in accordance with the Principle of Informative Changes and with the Principle of Perceptual Organization, specifically, the grouping law of similarity). In the key, the patches should appear in the same order as the content elements in the graph, presenting the key elements in a column or a row; if you use a column, make the top element refer to the leftmost content element (the bars in this illustration), make the second from the top element refer to the content element second from the left, and so forth. These guidelines are followed in **do**. (If your audience does not read from left to right, modify this recommendation accordingly.)

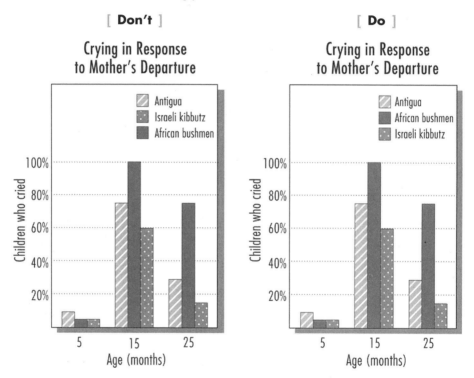

Figure 7.27. Readers will have to search for corresponding bars if the order of the key does not match the order of the content.

Use the content of one divided bar in multipanel displays to serve as the key.

If the segments of divided bars are ordered and colored or shaded the same way, labels at the far left of the display may be sufficient to identify the corresponding segments of all of the bars, as in **do**. This practice is particularly effective if two conditions are met: (1) The segments are approximately the same height—in this case, the horizontal orientation of the labels will lead the eye across the corresponding segments of the different bars, grouping with them via good continuation; and (2) the bars are close enough to group, so it is clear that the labels do not apply only to the leftmost bar.

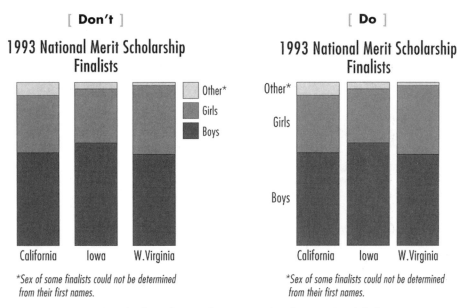

Figure 7.28. The first of several corresponding bars and its labels can serve as a key for the entire display—if the bars are close enough to group.

Creating Multipanel Displays

Multiple pie graphs or multiple divided-bar graphs can be used to illustrate different levels of two independent variables. For example, two pie or divided-bar graphs might represent the division of the work force into different jobs in two different years. Multipanel displays are also useful when you want to display more than one graph of the same format or more than one graph of different formats. The individual displays or panels should be produced in accordance with the recommendations for that type of graph. To create a display with two or more graphs, two major issues must be resolved: Which data should be presented in which panel? How should the panels be arranged in the display? If data answer different questions or your point focuses on specific comparisons, follow the first recommendation below; if all comparisons are equally important, skip the first recommendation and follow the remaining ones in this section.

Assign data that answer different questions to different panels.

Group data so that the reader can see the important trends and interactions, as is illustrated in **do**. As a first step, you may find it useful to state concisely (in writing or aloud) which comparisons are most important to convey. Ideally, you will help the reader to make important comparisons by plotting those data together.

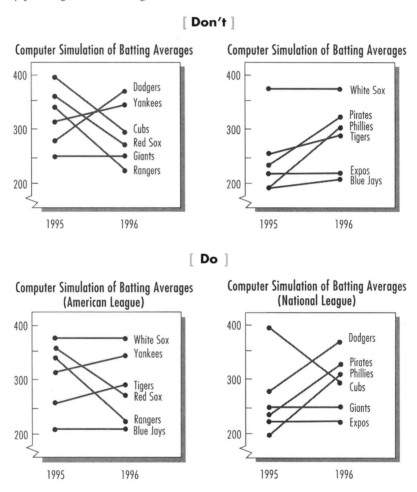

Figure 7.29. Data should be organized to help readers make specific comparisons.

Assign lines that form a meaningful pattern to the same panel.

Much of the power of line graphs arises from the fact that readers learn the meanings of specific patterns of lines. In most cases, assign lines that form a common pattern (e.g., parallel lines or an X) to the same panel, as in **do**. The exception: As noted above, if your point requires the reader to compare specific pairs of lines, graph those data in the same panel.

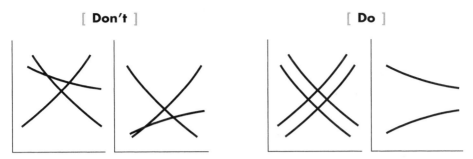

Figure 7.30. A judicious selection of data to be graphed in each panel may produce interactions that help tell your story.

Plot similar lines in the same panel.

If there are no comparisons of particular interest, and no meaningful patterns in the configuration of content elements, then plot similar lines in the same panels, allowing them to be grouped perceptually into relatively few units.

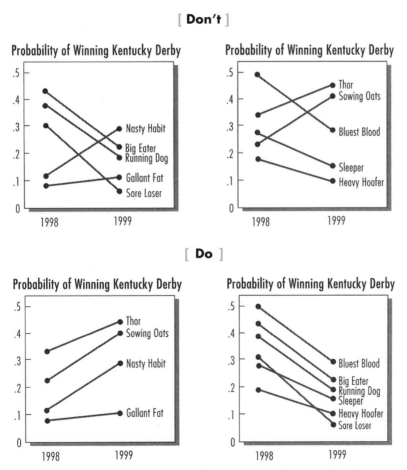

Figure 7.31. Picking a winner is difficult enough—the principles of perceptual organization can help.

Place panels as close together as possible without causing improper grouping of components.

Multiple panels should cohere into a single display; if they are too far apart, they will appear as unrelated graphs. If they are too close, however, our visual systems may improperly pair content elements or labels of one display with those of the other (in accordance with the grouping law of proximity). Put the panels as close as possible without seeming to touch or otherwise running the risk of improper grouping of labels or segments. Compare the spacing of panels in **don't** and **do**.

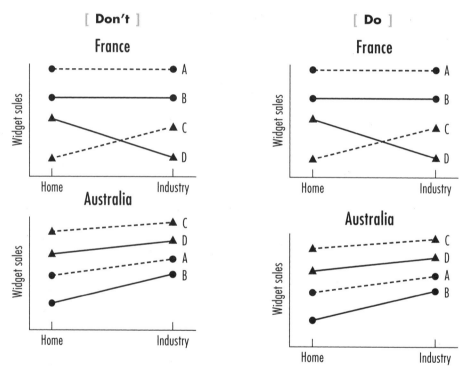

Figure 7.32. If panels are too close, labels and titles can easily be confused.

Put the most important panel first.

If one of the panels displays information of primary importance, as in **do**, put it at the left or, if there is more than one row, at the upper left. Given our reading habits, this is the panel that will likely to be viewed first. If your audience does not read from left to right, modify this recommendation accordingly.

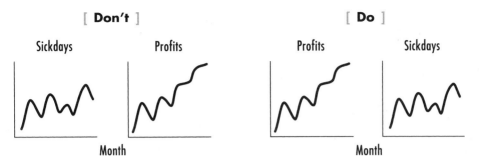

Figure 7.33. In most companies, profits are of more concern than are sick days.

Vary the appearance of corresponding content elements only to convey information.

According to the Principle of Informative Changes, all panels should be the same size and have the same general appearance unless size or appearance is used to convey information: Two different-sized pies might illustrate the relative amounts of weapons produced in each of two countries, the larger pie indicating that the corresponding country produced more weapons overall (but only a general sense of a difference can be conveyed by altering size in this way; we do not see area accurately). Or, a larger panel might be used to emphasize the main message. If such techniques are used, the title or caption should explain the point of the variation.

In addition, regions should be displayed in the same relative locations and in the same order in different panels, and regions that stand for the same thing should have the same appearance. If a dense cross-hatch is used for the "tanks" wedge of a pie graph illustrating the types of weapons sold by the United States, that pattern should be used for "tanks" in the pie graph illustrating types of weapons sold by France. Aligning wedges the same way facilitates comparing them. Hollands and Spence (2001) provided solid support for this advice to align corresponding wedges of multiple pie graphs. In an interesting twist, these researchers also showed that the relative advantages of pie versus divided-bar graphs could reverse, depending on the relative discriminability of the portions.

Use the same units along the Y axis in multiple panels.

When the same dependent variable is used in the different panels, space the scale units at the same distances along all Y axes, with the same number of ticks per interval.

Ensure that statistically significant differences appear different.

The Principle of Informative Changes implies that using identical scales on the different frameworks generally is best, but the Principles of Relevance,

Compatibility (more is more), and Discriminability imply that you should start the axes in different places to ensure that statistically significant differences look different. This advice is consistent with the recommendation in chapter 3 regarding when to truncate the Y axis: What the reader sees should reflect what the data show. As noted in chapter 3, if you truncate the axes, be sure to indicate this clearly by inserting breaks at the bottoms of the Y axes (e.g., inserting a zigzag or two slashed lines marking a gap at the bottom of each axis).

Overall heights in different panels should reflect overall amounts.

At first glance, from the **don't** version it seems as if all the countries spend roughly comparable amounts per pupil—but this is not true. The Y axes were excised so that significant differences could be seen between the pairs

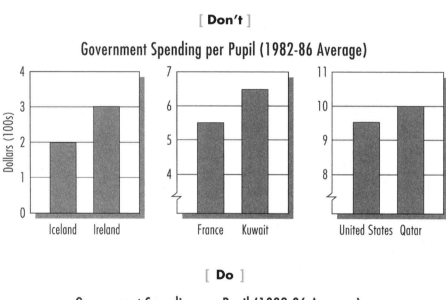

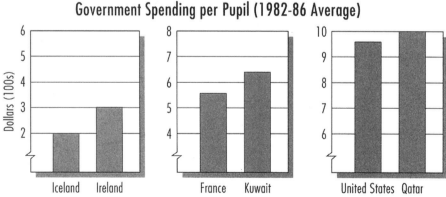

Figure 7.34. Arrange data to illustrate both pairwise comparisons and also relative overall levels.

of bars in each panel. Now look at **do**, in which the data were plotted so that the overall differences are also apparent. When the Y axis of one or more panels is not continuous, the Principle of Compatibility implies that the scales on the Y axes should be constructed so that the relative ordering of total amounts is preserved by the relative ordering of the overall heights of the content elements, conveying the correct visual impression of ordinal amounts.

However, if the overall amounts are critical, you probably should use the identical scale in each panel. DeSanctis and Jarvenpaa (1985) found that graphs with different maximum values on the scales were more difficult to use in a decision-making task than were graphs with identical scales; unfortunately, their graphs with different maximal values also did not use round numbers as scale values, which makes the result difficult to interpret.

Delete redundant labels on the axes if precise values are not important.

Both versions of the example illustrate the percentage of people that in 1985 and 1986 believed that nuclear war and terrorism were the most important problems facing the United States. The don't display is the more cluttered version and contains three redundant labels: "Year," "Percent," and the Y-axis numbers. If you divide data between two displays that are presented side by side, you often can center a single label beneath both X axes. It may also be possible to eliminate the labels of the Y axis of the right panel and move the two panels closer (so that they group more tightly), as is illustrated in **do**. However, keep in mind that eliminating the labels of

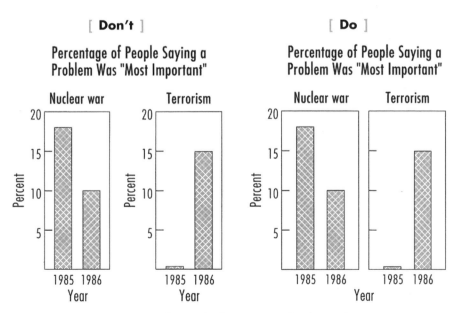

Figure 7.35. Every mark in a display competes for the reader's attention—eliminating unnecessary material is a good idea.

the Y axis of the right panel makes it difficult to obtain precise values, even with an inner grid; this practice is appropriate only if the point is simply to illustrate trends and interactions under different conditions.

Specify part/whole relations explicitly.

The top panel of the example shows stock prices over most of 1990; the bottom left panel shows the trend over the last week; the bottom right panel shows the trend over the last day of the last week. The arrows connecting the three panels are critical; they show how the panels interrelate. The titles of the individual panels are also essential. If a panel presents a second version of information in another panel, the relationship should be established by arrows or other visual means (e.g., a drawing of a magnifying glass for a detailed subpart of a function or an exploded section). Without such cues, the proper grouping will not be established.

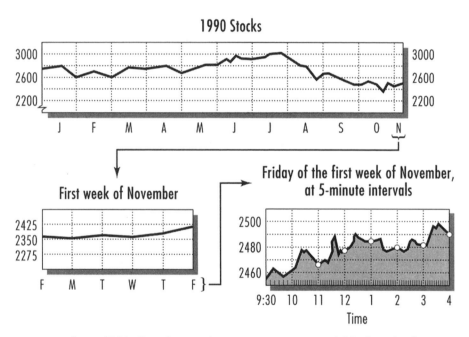

Figure 7.36. Complex graphs are sometimes unavoidable if much information must be presented. Indicating the relations among panels clearly helps the reader to comprehend the entire display.

**In multiple side-by-side graphs, put labels of pairs
at the far left and align the corresponding pairs
in different panels.**

When multiple side-by-side bar graphs are used, eliminate redundancy
by using only a single set of labels, as in **do**. Because we read from left to right,
it is best to put the labels at the far left so that they will be seen immediately.

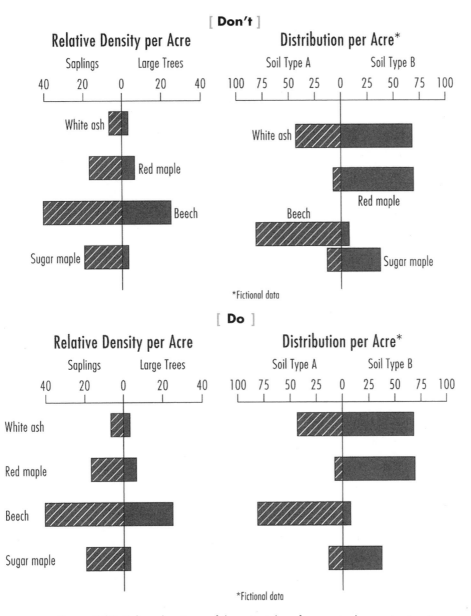

Figure 7.37. Take advantage of the principles of perceptual organization to
reduce the number of labels.

Ensure that the corresponding bars are aligned so that the good continuation serves to group each label with the corresponding pairs of bars in each panel.

The Next Step

If you have created a pie graph, divided-bar graph, or visual table, now go to chapter 3 to label its elements appropriately and provide a title (pages 95–105). For all other types of graphs, go to chapter 8. Many complex displays can only be created by an iterative process, by trial and adjustment. Sometimes the overall configuration of the display has unfortunate implications. An insignificant difference can appear visually dramatic or a real difference among measurements may not be obvious in the display. Chapter 8 provides recommendations that can be used to check how accurately the visual impression conveyed by a display in fact reflects the actual patterns in the data.

How People Lie
With Graphs

8

VISUAL DISPLAYS ALLOW us to gain a quick impression of relationships in a set of data. The Principle of Compatibility leads to a simple dictum for the display of quantitative information: More is (indicated by) more. Greater amounts of the measured substance should be depicted by greater visual amounts, either in extent (the height of a bar, point, or line) or in area (the size of a wedge in a pie graph or of a sector in a layer, divided- or stacked-bar graph). In this way, we can compare relative amounts at a glance:

Bigger visual differences are interpreted as reflecting bigger differences in the data.

This general idea is central to virtually all good graphical representations of quantitative data. As a designer of graphs, you use it to invent new formats. As a reader of graphs, you use it to interpret a format that is new to you: Simply look for the visual dimension or dimensions that are being varied and discover how those dimensions reflect variations in the content. Displays that do not use this technique are difficult to learn to use. For example, Chernoff (1973) invented cartoonlike faces that can have different overall shapes and different types of eyes, noses, and so on, where each value of each dimension (e.g., each type of nose) corresponds to a different value of a dependent variable. The reader must memorize by rote the arbitrary associations between the visual variations and what they represent, and later must recall these associations by rote.

But our immediate apprehension of "more is more" is a two-edged sword. It makes it very easy for us to interpret most types of graphical displays; it also enables a designer to deceive us easily, as was pointed out by Derrell Huff in his classic book *How to Lie With Statistics* (1954). To fool a reader, the designer merely needs to make the graph look as if there is more extent or area of one visual element than of another, and the reader will automatically assume that the actual entities represented vary in the same way. Many of the same techniques used to highlight important information or make actual differences visually obvious in a graph can also be used accidently or dishonestly to make spurious effects look significant.

To some extent, what counts as "misleading" depends on the point being made and the context. However, it is never appropriate to make a difference that is not statistically significant appear as if it were real. For example, if you measured extrasensory perception (ESP) ability in five men and five women and found that women scored one percent higher on the test, this difference probably would not be statistically significant; if an apparent difference is not statistically significant, it merely reflects the random variation in the population. Thus, it would be inappropriate— whatever your point or context—to graph the data to make this bit of random fluctuation appear as if it reflected something real about men, women, or ESP.

This chapter is not intended to be a tutorial in how to produce dishonest graphs. To the contrary, my intention is to show the principles underlying such misrepresentation so that the reader will recognize a deceitful display for what it is. If you have unintentionally produced such a display, this chapter will help you to catch the problem before it's too late. Moreover, to the extent that others may intentionally produce deceptive displays, the material in this chapter may help to discourage them: If enough people become aware of these principles, misguided designers may decide to stop using them to distort information.

The examples provided here do not exhaust the ways that visual displays can be used to lie, but they convey underlying techniques that can be used: alterations of the framework, distortions of the content, and variations in the design of multiple panels. The distortions illustrated here range from the subtle to the blatant. Let us see how people can lie, and be lied to, with various types of displays.

Variations in the Framework

Visual differences in bar and line graphs can be exaggerated or minimized in various ways. Some of these methods require altering the axes. Consider, for example, a series of four graphs illustrating the numbers of strategic warheads for the United States and the U.S.S.R. in 1991. The first graph has not been distorted, and two things are apparent: First, both sides have a lot of warheads; second, there is only a small difference in the size of the two arsenals. Now, compare this graph with the next three, which might be used by (surely hypothetical) opposing political candidates. (I treat the X axis and the Y axis as corresponding to the independent and dependent measures, respectively, but the reader should keep in mind that these roles are reversed if a horizontal bar graph is used.)

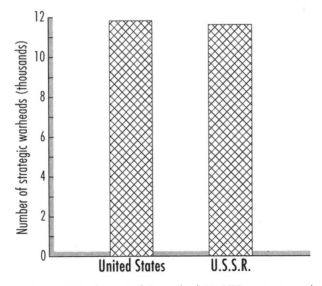

Figure 8.1. In 1991, the United States had 11,877 strategic warheads and the U.S.S.R. had 11,602. These data are presented in distorted form in the three following graphs. In order to focus attention on the competition between the United States and the U.S.S.R., in this series the labels for the two countries have been emphasized.

Do not excise the Y axis in order to exaggerate a difference.

Excising part of the Y axis is a useful way of making important variations in lines or bars noticeable. However, this technique, used improperly, can mislead the reader. The candidate arguing against a military buildup replotted the data on a graph whose vertical axis starts not at zero, but at 11,500. The difference in the number of warheads looks bigger on the page, and the reader is left with the impression that there in fact is a bigger difference in the stockpiles. Notice also that the excision is signaled by very small slash lines, which may not be detected by the casual reader.

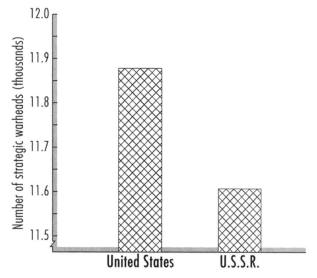

Figure 8.2. The data replotted using a Y axis with a restricted range of values (most of the scale has been excised), which greatly exaggerates the visual impression of the difference.

Do not provide a spurious range of values to minimize a difference.

Another candidate, arguing for a military buildup, might want to minimize the difference between stocks of warheads and draw the graph differently. By extending the scale on the vertical axis from 0 to 1,000,000, the designer has made the visual difference in the number of warheads very small indeed. To make the added range seem necessary, a spurious reference line may be included, such as the number of warheads that would assure a total nuclear winter if all were detonated; this added information serves to distract the reader from recognizing the psychological effect of the range used.

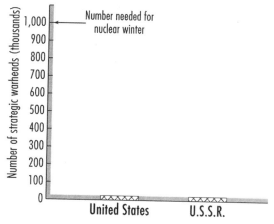

Figure 8.3. The data replotted using a Y axis with an expanded range of values (needed to include the number of warheads that would, by the most conservative estimates, ensure a total nuclear winter), which diminishes the visual impression both of the absolute numbers of warheads and the difference in numbers in the two countries.

Do not use a scale that creates a misleading impression.

Another way to minimize the visual impact of a difference is to alter the scale. For example, a designer could use a logarithmic scale on the Y axis. In this sort of scale, the distance between numbers becomes increasingly

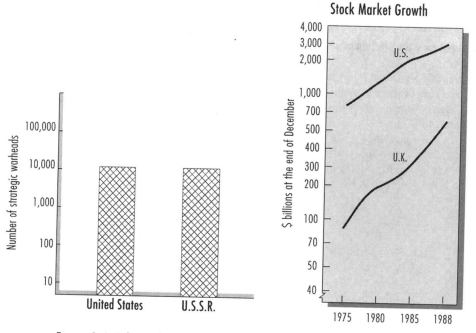

Figure 8.4. A logarithmic scale on the Y axis minimizes visible difference if, as here, the overall amounts are relatively large.

compressed as the numbers increase (see pp. 88–90 and appendix 1)—as the total amount increases, it takes larger differences in values to produce the same visual effect. This technique, used in the warheads example, erases any distinction between the stockpiles of the two nations.

Logarithmic scales are useful if the measurements have such a large range that critical differences among relatively small numbers are not visually evident—but the technique can be especially misleading if the content has two or more levels on a parameter (two or more lines, or two or more bars, per location on the X axis), as in the graph of stock market growth. In this example, the top line appears to show a smaller increase in stock market growth in the United States than in the United Kingdom, illustrated by the lower line; in fact, the increase in the U.S. stock market was well over four times larger than that in the U.K. market.

In addition, scales can be altered in other ways. For example, Cleveland, Diaconis, and McGill (1982) found that people overestimate the values of correlations in scatterplots when the data are plotted on scales with a larger range of values (which causes the point cloud to appear denser). Lauer and Post (1989) extended this line of research and not only showed that people tend to inflate the correlation between two variables when scales are altered so that the visual density of the points increases (i.e., they appear less dispersed), but also do so when greater numbers of points are plotted, or when the slope of the plotted line makes the points group closer to the best-fitting line. Lauer and Post point out that most graphing programs adjust the scales of a graph on the basis of the range of values to be displayed, which often results in different scales for different sets of data—which in turn will distort our perceptions of correlation to different degrees. In spite of such distortions, Lauer and Post also found that, in general, people tend to underestimate the correlation of data shown in a scatterplot.

Moreover, Meyer and Shinar (1992) found that teachers of university-level statistics were no better than high school students in avoiding illusions that affect the perception of correlations from scatterplots. Both groups were affected, and to comparable degrees, by visual properties (e.g., scale lengths) that affect our perception of correlation.

Do not vary the aspect ratio to mislead.

We interpret a steeper line, or a sharper increase in the height of bars, as reflecting a larger increase in the measured quantity. A dealer trying to sell Asian elephants to a zoo might use **don't** to imply that Asian elephants are considerably smaller than African elephants and thus have appreciably lower upkeep costs. The **do** version presents the same data, but the difference by species is not nearly so dramatic because the slope of the line more accurately reflects the actual difference between the average weights of the two species. These graphs show how variations in the aspect ratio—the ratio of height to width—can alter the visual impression of a difference. The effect of increasing or decreasing this ratio is to make the rise seem steeper or less precipitate. These visual impressions can be mistakenly translated by the reader as meaning a greater or lesser difference in the measured quantity. (For evidence that visual impression is in fact altered by aspect ratio, and a quantitative discussion of the degree to which aspect ratio and other factors affect one's impression of how steeply a content line rises, see Simcox, 1984.)

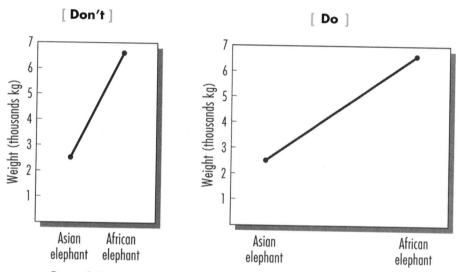

Figure 8.5. Increasing or decreasing the heights of a display relative to the width can visually—and inaccurately—exaggerate a difference.

Do not vary the starting places of bars.

The relative height of a bar or line conveys amount; the graph will be misleading if the baselines of these elements vary within a display, thereby changing relative heights. This effect is employed in "Saudi Arabian Oil Production," where the graph is inventively drawn on the back of a camel. Prices did not in fact peak in 1986 (they were the same in 1986 and 1988), nor did they rise as sharply in earlier years as it appears here. The appropriateness of the camel illustration may distract the reader from noticing that the designer has added overall height to some bars by increasing the height of the base.

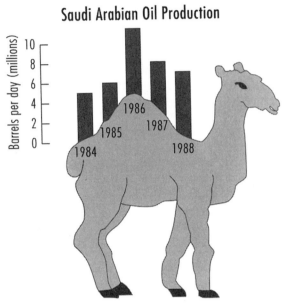

Figure 8.6. The lengths of the bars, not the heights of their tops, convey the information in this display (which is more like a visual table with a scale than a graph); because the baseline varies, the tops of the bars do not indicate the amounts.

Do not add a spurious reference line to alter perceived differences in height.

The designer can also include a reference line to alter the perceived origin. Now the reader sees the tops of the bars or lines relative to the heavy horizontal line. The example illustrates the 1985 energy consumption in tons of coal-equivalent in the United Kingdom, West Germany, and Japan. The reference line is the energy consumption of Canada (which has nothing to do with the topic); its inclusion visually amplifies the difference between the United Kingdom and Japan. The effect of a spurious reference line can be emphasized by shading or coloring the portions above the line, as was done here. Recall that we pay attention to differences in visual properties: We note proportions, not absolute amounts. Thus, by visually defining shorter bars (the portions above the reference line), the designer makes the differences in the heights of the tops more striking than if the entire lengths were compared.

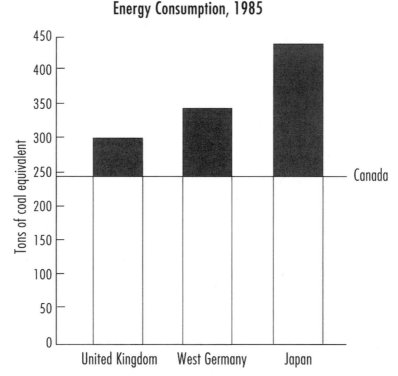

Figure 8.7. The eye notices the dark portions of the bars, which have greater proportional differences than do the bars overall—and thereby the reader is misled into seeing larger differences than are actually present.

Do not use a three-dimensional framework to exaggerate sizes.

Many displays are drawn in perspective, as if they were three-dimensional. This technique provides many opportunities, seized on by some designers, to fool the eye and the mind. The underlying phenomenon is known as size constancy, the tendency to perceive an object at its actual size at whatever distance it is seen, even though the object would occupy a smaller region in a photograph when it is farther away. A distant car does not look smaller than a near one, it just looks farther away. So, if two pictures of cars are drawn the same size on the page, but one is drawn so that it appears to be farther down the road, away from the viewer, that one will be seen as the larger of the two. By making the framework look three-dimensional, a designer can take advantage of this predilection to distort the perceived sizes of objects. Below, this technique has been used to diminish the visual impression of the difference between Qatar and the United States in spending per pupil. The bars "farther away" look larger than they should because the visual system compensates for the apparent distance. Note that the baseline is different for each bar, which also contributes to the impression of increased height.

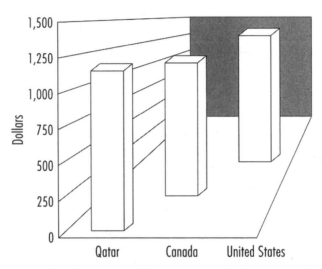

Government Spending on Education per Pupil

Figure 8.8. The phenomenon of size constancy leads us to see the farther bars as larger than they are, an impression that is reinforced by the added height on the page of the more distant bars.

Do not use perspective to change the scale at different locations.

A framework drawn in depth is sometimes used, perhaps inadvertently, to change the scale at different locations along the X axis. In "Oil Revenues," the graph is drawn on a drum, and the display would suggest that oil revenues peaked in 1985 and declined rapidly thereafter. Such a display might be included as part of a plea to raise prices. Because the central part has the most vertical range and both sides are compressed, this graph distorts the relative extent of the differences found in the middle years. The reverse effect will occur if the display is drawn on a surface that curves toward us (e.g., the inside of a coffee mug, seen from over one edge of the rim); when we look into a concave display, the middle portion appears the farthest from us. Designers sometimes produce graphs that are drawn on a wall that is receding into the distance; because the range of variation is expanded for the content on the "nearest" parts of the wall, these differences will be emphasized.

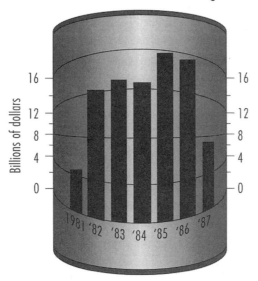

Figure 8.9. The scale at the middle is exaggerated compared to the scale at the sides, which exaggerates differences among bars near the center relative to those at the sides.

Variations in the Content

The data themselves can be transformed to produce an inappropriate visual impression. Special tricks can be used with specific sorts of content material, namely, bars, lines, and regions.

Do not use percentages to exaggerate or minimize changes.

A particularly compelling example of the visual effects of transforming the data is illustrated in the two stock market graphs. "Size of Stock Market" tracks the absolute size of the stock market in three countries over three years. In the "Percent Change" graph, the absolute amounts have been converted to percentage increases over the previous measurements. Note that the relative percentage of increase is larger for Japan than for other countries, a fact not immediately obvious in the first graph. If a designer wanted to emphasize Japan's economic strength, "Percent Change" could be used to conceal the fact that the United States actually had a much bigger stock market during these periods. It is no mystery which version would be used to sell U.S. bonds.

Moreover, one can choose what percentage to graph: percent over all countries, percent relative to some baseline (overall average, lowest profits, highest profits), percent relative to the same country at an earlier time. Different visual effects will arise depending on what numbers are plotted, and numbers can be chosen that convey almost any desired effect. The use of percentages may be appropriate—but not if they obscure information the reader needs to understand the actual data or to make specific decisions.

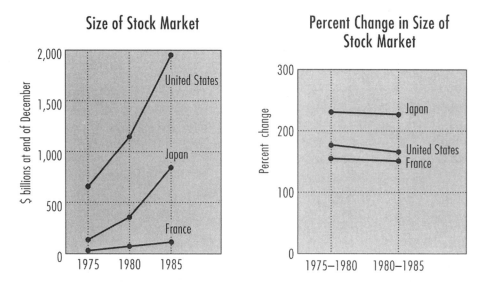

Figure 8.10. The U.S. stock market had larger increases per year and was much larger than Japan's (left); graphing percent change (right) obscures those facts.

Do not use derivatives to mislead.

Another transformation that can cause visual displays to tell one story when the data in fact tell another hinges on the use of the mathematical concepts of first and second derivatives. The first derivative is the rate of change, the second the rate of change of this rate: In the example usually given, the first derivative of distance with respect to time is speed, the second is acceleration. One might graph the rate of population increase using a first derivative, which would indicate the increase from year to year in the number of people being born; the second derivative would indicate the rate at which the size of the increase changes. The second derivative can be very large even if there is only a modest increase in actual amount relative to a small base. Many readers, unfamiliar with derivatives, will interpret the bars or lines as displaying absolute amount. Even for mathematically aware readers, such a transformation can effectively obscure embarrassing differences (or lack of differences) in the absolute amounts.

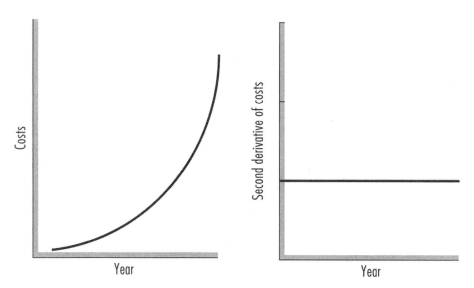

Figure 8.11. The graph at the right indicates that the rate of change in the rate of change was constant—but does not specify what the rate of change was. As is evident from the graph on the left, very large rates of change can be constant.

Do not graph difference scores to mislead.

A closely related technique is to graph only differences between two or more measurements. In this case, the dishonesty arises when the designer chooses an inappropriate number to subtract from each of the amounts. The example shows the stock market data from page 212, plotted in a different form. Here the display presents the difference in the percent change from the two intervals (constructed by subtracting the left number from the right number plotted in the "Percent Change" graph for each country) and conveys a very different impression of the economic status of the three countries.

Difference in Percent Change in Size of Stock Market, 1975–1980 vs 1980–1985

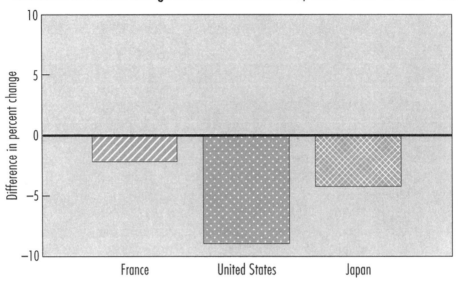

Figure 8.12. The reader has no way of knowing from this graph that the U.S. stock market in fact was the largest and fastest growing of the three.

Do not vary the thickness of wedges.

Below, the intent is to emphasize the growing maize production of China. The wedge for China has been exploded from the rest of the pie. It has also been drawn thicker, so its overall volume is larger—and it appears to represent more than it actually does. This effect depends on the Principle of Perceptual Organization (specifically, the aspect of integrated vs. separated dimensions): You must expend great effort to pay attention to width without also paying attention to height, and usually will take in the two dimensions together, combining them into the impression of overall area or volume.

Maize Production

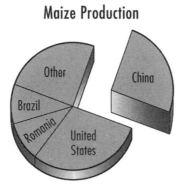

Figure 8.13. Making the exploded wedge thicker makes it seem to represent more.

Do not use three-dimensional edges to exaggerate the apparent size of the content.

The "Maize Production" pie graph has been drawn with a bit of a lip added in front to suggest a three-dimensional effect. This add-on carries the rim along quite nicely—increasing the visual impression of the U.S. share of total maize production.

Maize Production

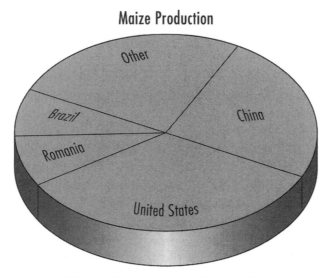

Figure 8.14. Adding the three-dimensional lip at the front exaggerates the size of the wedge for the United States.

Do not use brightness of color to exaggerate area.

Compare the two versions of "Apple Production." Brighter areas are perceived as larger than darker ones; this effect, called *irradiation*, is a special case of the Principle of Compatibility (specifically, the aspect concerned with perceptual distortion), and applies to brightnesses of color and, to a lesser extent, shades of gray. Using this variable to manipulate salience or to convey quantitative differences must be done with care to ensure that the display does not mislead.

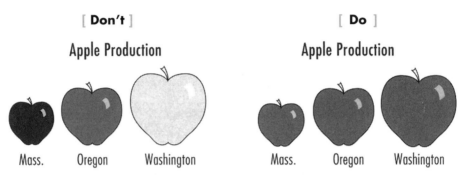

Figure 8.15. Adding intensity exaggerates the visual impression of the magnitude of the differences in apple production.

Do not use areas to minimize differences.

In pie graphs and divided-bar graphs, the area of each sector conveys the relative amount. These displays can mislead if the visual impression of area is distorted. It is especially easy to be misled by such displays because our visual system is poor at estimating area; we systematically underestimate areas as they increase in size. So if one uses relative areas to convey amounts, one can deemphasize an increase.

Do not covary height and width.

All three versions of the following example purport to illustrate the brain weights of different animals. The top graph is not distorted visually, although for purposes of correlation of brain size with intelligence, the data themselves are misleading; the appropriate measure would be the ratio of brain weight to body weight, and cows would appear far less relatively sentient. The middle display might be used by someone who is opposed to the eating of red meat. This graph, by the selective use of a wider bar, seems to suggest that cows have larger brains than gorillas. The bottom display might be used by someone who favored meat eating. When participants were asked to evaluate the rate of apparent increase from left to right, they rated the middle display as increasing more sharply than the top one, but the bottom display as increasing less sharply than the top one (Kosslyn & James, 1980). Height and width are integral perceptual dimensions: Readers do not immediately see the height or width per se; they see the size of the rectangle. This fact provides additional possibilities for lying with bar graphs and step graphs by varying the widths of the bars or steps as their height is varied.

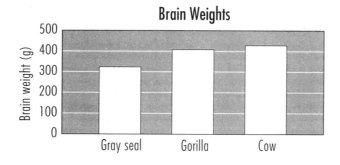

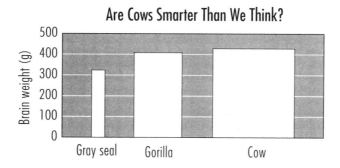

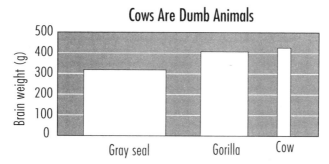

Figure 8.16. Varying the thickness of the bar manipulates the immediate visual impression; the middle panel sends the message that cows are brainier than the other animals.

Do not vary the weight of shading, stripes, or cross-hatching to mislead.

The eye is drawn toward more visually salient regions. The person opposed to eating red meat might be tempted to create a graph like **don't** to draw the reader's eye toward the largest bar; making it more salient emphasizes its role in the display and leads the reader to overestimate the size of the increase (Kosslyn & James, 1980, found this to be a rather small effect, however).

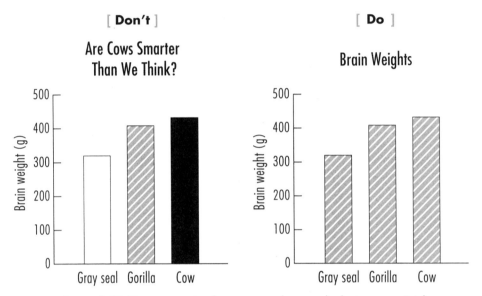

Figure 8.17. The progression from gray seal to cow looks inappropriately sharper when additional ink makes each bar increasingly salient. If the background were dark, or if the display were on a computer screen, adding more lightness would have the same effect.

Do not use occlusion to make bars look larger.

Tricks used to produce three-dimensional bars also can mislead the viewer. The example illustrates the percentage of people in each of five income groups. The rearmost bar represents 32.5% but looks larger than that—a useful effect if the designer is trying to convince people to keep in office the incumbents who created so much prosperity. We see partially covered objects as farther away than the objects that cover them. By having the bars partially overlap, the mechanisms of size constancy are brought into play, so the rearmost bars seem larger than they are.

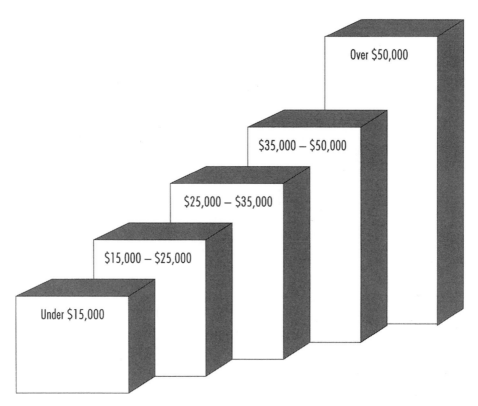

Figure 8.18. We see occluded objects as behind the occluding ones, and size constancy operates to make the occluded objects appear larger than they otherwise would.

Do not vary width within a line.

The actual slope of "Average U.S. Airline Fares" is a line that bisects the exhaust of the plane—but the perceived slope can be altered by shading the top or bottom of the exhaust, as in **don't**. If the lines that form the content of a line graph vary in width as shown here, the slope will be exaggerated if the bottom edge is emphasized, or minimized if the top edge is visually stressed. This technique looks silly unless the designer, as here, found a rationale for decorating the content line.

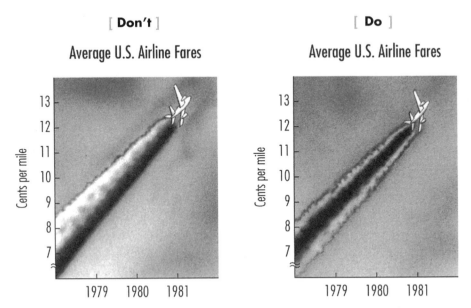

Figure 8.19. In the graph on the left shading has been used to emphasize the trend, making it seem steeper than it is. Shading the top would de-emphasize the trend.

Do not vary depth to distort line height.

If the content line in a line graph is drawn as a three-dimensional bar, one end can appear closer than the other. The height above the Y axis (at the "front" of the display) can be exaggerated if part of the line is drawn to appear relatively far away. In the example, the increase in the German share of world exports is exaggerated because the right portion of the bar appears higher than it should. By having the bar snake back, the designer makes the right part actually higher on the page—and the ambiguous depth cues do not allow the reader to have a good sense of how high the bar is above the "floor." This sort of exotic display is most convincing if the line is drawn as an object (a bar of steel, in this case) that makes sense in context.

German Share of World Exports

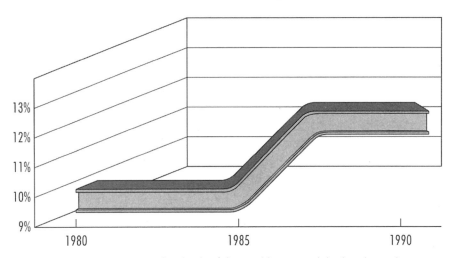

Figure 8.20. Varying the depth of the steel bar varied the height on the page, making it very difficult to read the actual values off the Y axes.

Do not insert a line in a scatterplot by eye to emphasize a trend.

When a cloud of points is presented in a scatterplot, fitting a line through it will enhance the visual relationship between values of the variables along the X and Y axes. In **do**, there is in fact no systematic relationship between the X and Y variables, but by putting differently shaped lines through the cloud (flat, U, increasing), we get an impression of such a relationship. The

[**Don't**]   [**Do**]

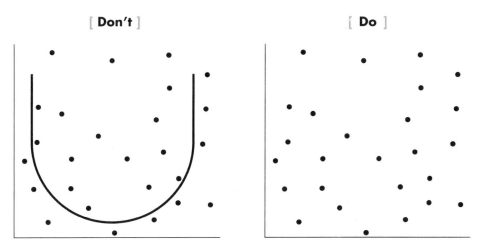

Figure 8.21. A line can impose a pattern on the points because they group with it (via proximity and common fate). As is evident in the graph on the right, there is no actual systematic pattern of variation in the plotted points.

effect occurs because of both the grouping law of proximity (points near the line are grouped with it) and the law of common fate (points lining up with the line are grouped with it). This sort of manipulation only can be regarded as an intentional effort to mislead the viewer. The practice is particularly dishonest because there are well-established statistical techniques for fitting lines to data (see appendix 1), and most readers will assume that such procedures were used to place the line.

Do not vary the aspect ratio to make dots in a scatterplot closer to the lines.

Varying the aspect ratio has the effect of making the points in a scatterplot seem nearer or farther from a line. When the X axis is relatively short, the points seem to fit the line better. The graph in the example shows the amount of time participants in a study needed to scan across a visualized map with their eyes closed. Although it is clear in both versions that more time is required to scan a farther distance, the point is made more dramatically in the version on the right. This manipulation creates the impression of a stronger relation between the independent and dependent variables if the width is decreased, or a weaker relation if the width is increased.

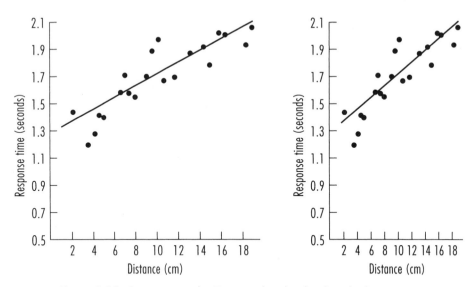

Figure 8.22. Compressing the X axis makes the dots hug the line more closely and thereby enhances the impression that they fit the line.

Multiple Panels

Another set of techniques that sometimes is used to mislead depends on breaking the display into two or more separate displays.

Do not vary the size of a panel to mislead.

The example shows the amount of time that first-graders, fourth-graders, and adults required to "see" large or small features of visualized animals (e.g., the back or the whiskers of a mouse) versus the amount of time that they required to recall the same features without visualizing them (Kosslyn, 1976). The large panel shows that adults required different amounts of time to recall the information in the two ways. Note that one of the small panels shows a different trend: First-graders tend to use mental imagery even when

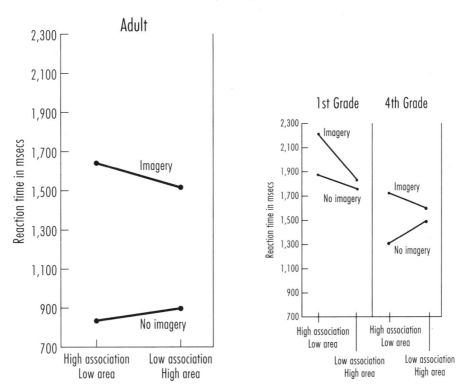

Figure 8.23. Making a panel larger and putting it on the left encourages the eye to linger on it—and thus possibly to neglect the inconsistencies to its right. This would be unfortunate here, because the difference for the first graders is what makes the graphs interesting.

they are not asked to do so. By drawing one panel of a multipanel display larger than the others, disparities can be—literally—minimized.

Do not use different transformations or scales to distort the visual impression.

If there is an embarrassing difference between two sets of comparisons, the designer may try to conceal it by using different transformations (perhaps converting the data to logarithms in one panel and to difference scores in another). The reader will find it very difficult to make direct comparisons. Sales of the weakest division of a company might be shown as the percentage of increase, which is to be compared with actual sales for the other divisions. Of course, the designer would offer a rationale for this distinction, such as the relative youth of the weakest division and the claim that the amount of early growth is most important.

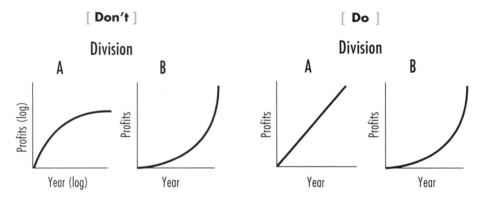

Figure 8.24. Supporters of Division B might try to slip by the display on the left, arguing that Division A profits are about to peter out. The reality, shown on the right, is that both divisions are winners.

Do not combine distortions.

The three versions of "U.S. Exports to Mexico" illustrate a few of the ways in which techniques might be combined to produce even stronger effects. All visually exaggerate the amount of growth in the period 1989–91 and might be used to discourage Mexico from entering into a free-trade agreement with the United States. Research has shown that the techniques are not strictly additive: Combining two equally distorting techniques does not make the display twice as distorted psychologically (Kosslyn & James, 1980). Nevertheless, such combinations do result in more striking—and thus misleading—displays.

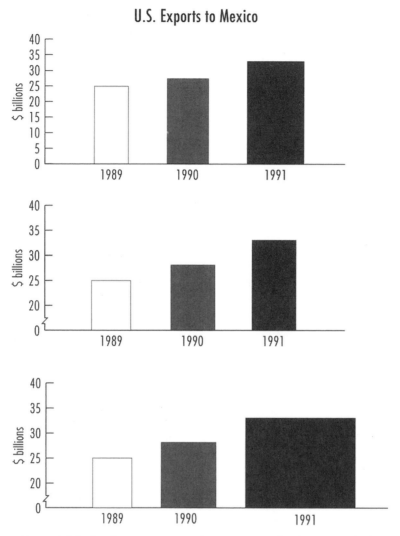

Figure 8.25. Combining various techniques varies the degree to which a trend or difference is seen to increase or decrease.

The
Next
Step

We have seen how psychological principles can be exploited in a variety of ways to distort the reader's impression of a graph's message. You should take care not to use these techniques inadvertently to mislead the reader, and you should be alert for the use of such techniques by others (and the ones shown here are not the only ones): They are surprisingly pervasive in the media. If a pattern in the data is not statistically significant, it should not be visually striking; if a difference or trend is obvious in a graph but is not actually significant in the data, the graph is misleading. Similarly, if a difference or trend in the data *is* statistically significant, it should be visually obvious. If you find that your graph is misleading, the next step is to correct the false impression by altering the framework or the content or both (without violating any of the relevant recommendations).

Beyond the Graph

THE RECOMMENDATIONS PRESENTED in this book are rooted in the recognition of a central fact about the reader, not the display: The reader, as a human being, has a particular perceptual and cognitive system with identifiable strengths and weaknesses. That being so, the principles discussed in the preceding chapters will be in force whatever the specific nature of the display—the eye and mind, in their intimate partnership, will group elements, will focus on salient visible differences, and will in general behave as described. The nature of the reader will not change, but the display the reader is looking at will not always be a graph. In this chapter I offer recommendations for using the principles of perception and cognition when creating other types of visual displays.

We can divide visual displays into two general classes: those that convey quantitative information and those that convey qualitative information. Graphs provide quantitative measures (e.g., the price of tea), even if the measurements apply to discrete entities (e.g., countries). Maps also fall into this category; as a consequence of portraying the layout of territory, they implicitly specify distances between locations, a quantitative variable. Often, they also specify the amount of something at each location, perhaps height above sea level, population, or annual rainfall. However, maps cannot be used generally to convey quantitative information; they are tailor-made to convey information about spatial layout.

In contrast, charts and diagrams are qualitative. Charts specify qualitative relationships among entities; family trees and flow charts are kinds of charts. Diagrams are schematic pictures of objects or events that rely on conventionally defined symbols (e.g., arrows to indicate forces); "exploded" drawings that show how parts of an object fit together and illustrations of football plays are diagrams.

However, other terminological conventions are sometimes used. B. Winn (1987) characterizes charts, graphs, and diagrams in a similar way to that offered here, but his "charts" can be entirely textual. According to the present characterization, a chart can include only text if spatial relations are used to convey additional information about the material. For example, an organizational chart might have one name at the top, two in a row directly under that, six under each of those names, and so on. If the reader understands that this is an organizational chart, it can be read as specifying a hierarchy. In addition, Moxley (1983) distinguished among three types of diagrams, some of which are charts by the present characterization; consistent with the present approach, Fleming and Levie (1978) emphasize the role of visual representations of relations (e.g., inclusion, subordination) in charts. Bertin (1983) offers a very elaborate discussion of such matters (see Wainer & Francolini, 1980, for a discussion of the rudiments of Bertin's ideas; see Kosslyn, 1985, for a review of Bertin's book). Tufte (1983, 1990) provides many instructive illustrations.

In general, if a display is intended to reach a general audience, it should rely on common, well-known symbols; Dreyfuss (1984) and Horn (1998) provide catalogs of such symbols. Tversky, Zacks, Lee, and Heiser (2000) summarize a set of symbols that people have apparently learned to interpret in consistent ways in Western culture. Specifically, they infer that (1) enclosed figures (including a standard bar, in a bar graph) suggest distinct sets of contained elements, (2) solid figures suggest discrete objects, even if their shape does not correspond to that object (hence a square can be used on a map to indicate a building that in reality has a complex footprint), (3) lines suggest connections (symbolically, as in family trees, or geographically, as in roads on maps), (4) crosses suggest intersections of paths (on maps), and (5) arrows suggest a direction or asymmetry in causality, time, space, or motion (as in flow charts).

In this chapter, I offer some specific recommendations for charts, diagrams, and maps. These recommendations, by no means exhaustive, are meant to underline critical issues you should consider when designing these displays. They also show how the present approach can be generalized to a wide range of display types; it should be easy to see how the eight principles can be applied to each type of display. Each of the following sections begins with a few recommendations about the circumstances in which you should use the format, and then turns to recommendations for designing that type of display. I finish with a brief look to the future of display design.

Charts

By some definitions, a chart is considered to be any display that has a symbolic content—it stands for something other than what is directly pictured (e.g., B. Winn, 1987). This general characterization would include some maps (navigational charts) and graphs. I use the term here in a more restricted sense: Charts convey not amounts but relationships, not "How many work here?" but "Who works for whom?" Tables of organization, flow charts, and family trees are all charts, visual displays that arrange information into categories or structures. In graphs, the sizes of elements correspond to amounts of the entity being measured (which is why I refer to pie *graphs*, not pie *charts*). In charts, size may not correspond in any way to the entity depicted, and if it does, size will reflect not amount but a quality of the relationship such as subordination or inclusion. Usually, lines connect the elements of a chart to indicate structure, but the arrangement of the units (often boxes) on the page may be enough to show the relationship clearly.

Use a chart to convey overall organizational structure.

Using a chart to convey information about the organization of discrete entities takes advantage of the Principle of Compatibility. Relationships such as "is a member of," "follows," "works for," and "is descended from" are illustrated clearly by the spatial relations of the entities in the display. A chart is a good idea if you want the reader to gain a sense of the overall organization of components, such as in the management of a company, the degrees of kinship in a family, or the sequential steps in a process.

Use a description to convey few entities and relations, a chart to convey combinations of comparisons.

The alternative to a chart is a verbal description, which is preferable if only a few entities or relations need to be considered. A simple statement, "The company has two components, consulting and publishing," is better than a tree diagram. A short description probably can be read faster than a chart can be deciphered, and it will tax short-term memory less than would a multielement display. On the other hand, if the reader needs to compare numerous different combinations of entities (e.g., the relations between various managers and their superiors in the company), a chart is better than a description.

More inclusive categories should be represented by symbols higher in the display.

The problem is to show how the X Corporation breaks down into separate divisions. The Principle of Compatibility leads us to place the superordinate entities higher in a chart, with the subordinate entities depicted beneath them (for related ideas, see Fleming & Levie, 1978; B. Winn, 1987). X Corporation has automotive, consumer electronics, and ranching divisions. These would be symbolized in a chart by boxes (or pictures of representative objects) in a row directly beneath the box (or picture) representing the company. Beneath each of these three entities would be boxes or pictures for their components.

Use a layout compatible with the materials.

The Principle of Compatibility (including the aspect of cultural convention) applies to charts as well as to graphs. A chart detailing the command structure of an organization should start at the top and work down, as in **do**. By convention, a sequence over time progresses from left to right, so a flow chart illustrating the steps of a manufacturing process should start at the left and work to the right. However, the convention in computer science is to start a flow chart at the top and progress downward. This convention was adopted because the lines of a computer program are listed down a page, and each successive component of a flow chart corresponds to a successive unit of the program's code—and thus the cultural convention is preserved within the microculture of computer programmers.

[**Don't**]

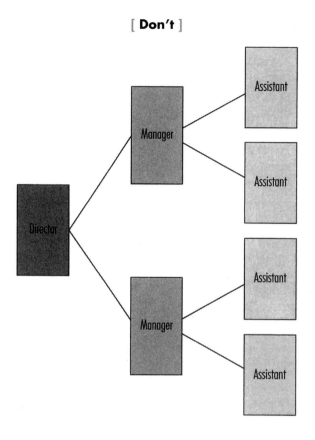

[**Do**]

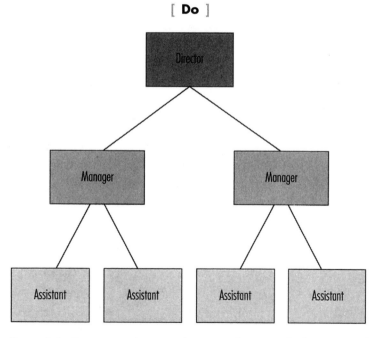

Figure 9.1. Category inclusion and command structure both correspond to vertical spatial relations; events over time are shown better in a horizontal layout.

Identify relationships.

Remember the muddled display of nutritional information presented in chapter 1 on page 20? Although the components of that display are two graphs and a table, the display as a whole functions as a chart. But, unhelpfully, the arrows leading from the center panel indicate different and unlabeled relationships. In keeping with the Principle of Relevance, identify every important piece of information in a display, as in the **do** version of this family tree.

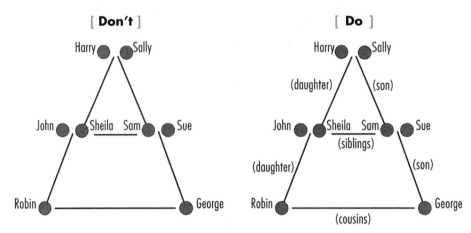

Figure 9.2. Every important piece of information, whether entity or relationship, should be identified in the display.

Use different colors, shading, or line weights to organize the components of complex charts.

The **don't** version of the following chart is difficult to take in; **do** is clearer. A chart is useful in conveying a complex set of qualitative relations, but the complexity must not visually overwhelm the relations depicted. This can happen if the number of entities or the number of relations among them is large. The spatial organization of the display, if properly produced, will allow the various aspects of perceptual organization to reduce the load on processing. To avoid producing a tangled web, divide the relations into two or more types and use different colors, shadings, or line weights for each: In **do**, the phases of the data collection process are distinguished from the final decision processes. If the content material does not lend itself to such a division, highlight what you consider to be the most important components. In some cases, however, even this will not help, and you should break the display into two or more individual displays.

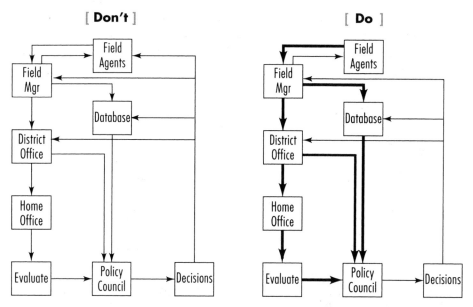

Figure 9.3. A complex chart can be sorted out relatively easily if it is visually organized into components.

Use familiar formats.

Charts can depict many types of qualitative relations and can do so in many ways. It is tempting to invent new and exotic types of display designs. This is well and good if the reader can quickly discern the entities and relations, but as any reader of national news magazines knows, this is not always the case. To be on the safe side, be sure that the reader is familiar with the conventions and types of display you are using.

Diagrams

Diagrams are pictures of objects or events that use conventionally defined symbols to convey information—to show the wiring of your kitchen, the nitrogen cycle, or the assembly of a model 1968 Mustang convertible. Diagrams combine literal elements (pictures of parts) and symbolic ones (arrows to show movement, direction, or association; shading to show curvature).

Do not explode an object too widely for recognition.

The brain processes shape and spatial relations in separate systems and does not always combine the two very accurately (which is one aspect of the Principle of Compatibility, specifically the aspect of spatial imprecision). Do not expect the reader to extrapolate or remember precise spatial relations among parts of a display. In an exploded diagram, show parts near their actual locations on or in the object depicted, as in **do**.

[**Don't**]   [**Do**]

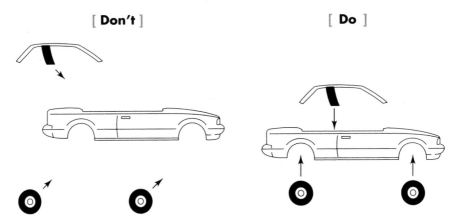

Figure 9.4. If spatial relations are distorted to show components, the distortion should be easy to reintegrate; otherwise, readers will have to expend effort to see the components in place.

Include only relevant material and ensure that it is visually salient.

Always respect the Principles of Relevance and Salience: The most important material should be highlighted, and unimportant material deemphasized or deleted. A diagram meant to show the driver's controls in an automobile may not be helpful to the reader if all automotive systems are drawn in, as in **don't**.

[**Don't**]   [**Do**]

Figure 9.5. A diagram intended to show the driver's controls is easier to read if the irrelevant components are eliminated and the relevant ones are highlighted.

Make conventions clear.

Certain disciplines—electronics, genetics, and linguistics, to take a few examples—have their own vocabularies of symbols, which should be respected. Depending on context, however, the same symbol may have different meanings: An arrow may indicate parts to be matched, direction of force, or movement. If you think your audience may be in any doubt about the meanings of the symbols you are using, state those meanings in a caption.

Maps

Maps are drawings that function as pictures of a physical layout: The features of a room, your town, the earth, the sea, and the sky can all be mapped. As stylized pictures of a territory, they provide information about locations and relations among them (usually in terms of relative distances and routes between them); by using conventional markings, they can also provide quantitative information about various locations, such as average temperature, voting patterns, and population distribution (for more information about when to use and how to design maps, see Fisher, 1982; Monmonier, 1991).

Use a map if more than one route is possible.

One alternative to a map is a verbal description of a territory. If the reader can reach the destination from a number of different starting places or by more than one route between two locations, a map is distinctly more helpful (remember the last time you asked for directions). A map is also preferable if the reader is likely to want information about the distance or spatial relations among numerous locations. In these circumstances, a map will reduce the load on short-term memory.

Use a map to label complex sets of information about a territory.

Maps are also a convenient way to label information about localities. A map allows the reader to see the relations among different values in many locations. For example, if population is presented (perhaps by bars standing on particular locations), the reader can not only see the relation of population to specific landmarks, such as rivers or the sea, but also gain an impression of the pattern of population variations as a whole. A map allows symbolic content to be grouped with locations via the grouping law of proximity.

In fact, maps and visual tables can be combined, creating a hybrid format. Cleveland and McGill (1984a), Cleveland (1985), and Dunn (1987) have argued that a "framed rectangle chart" is more effective than simply making regions of a map darker to indicate greater amounts at that location. A framed rectangle chart has a constant-sized bar (the frame) at or near the center of each location, and a bar is presented within the frame. The frame has horizontal ticks, allowing the reader to estimate the height of the bar.

Dunn (1988) tested the effectiveness of such displays. He first prepared maps that depicted murder rates in the United States; in one version, a framed rectangle was imposed on each state, and in the other version different amounts of fill were placed in each state. The participants could in fact read quantities from the framed rectangle display more accurately than from the filled map (such displays are termed *choropleth maps*).

We can extend the basic concept of a framed rectangle display, replacing bars with pictures. For example, in a map indicating population in different regions, the bars could be replaced by drawings of people sitting on each others' shoulders, with each person's representing a fixed unit amount of population. In such cases, the map serves to label the content elements of the visual table.

Ensure that features are easily identified by readers.

If the map includes specific features, such as particular buildings or landmarks, ensure that the icons are easily identifiable by the intended readers. Johnson, Stamm, and Verdi (2000, as cited in Verdi & Kulhavy, 2002) compared memory for features on maps that resembled what they represented to memory for features that were entirely symbolic. The features could be labeled or not labeled. The participants studied the maps, read a text passage and attempted to recall it, and finally were asked to reconstruct the map. The participants did much better when they studied maps in which features resembled their referents, as opposed to being indicated by arbitrary symbols, and labels also facilitated recall.

In addition, Lowe (1993) asked participants to study weather maps; half of the participants were meteorologists and half were not. When later tested, the meteorologists recalled more markings and did so more accurately. Having the relevant knowledge to interpret the maps allowed them to organize the material effectively and thereby recall it more accurately. Many researchers have found that people who have had experience using specific types of displays become more adept at using them (e.g., Barfield & Robless, 1989; Shah & Hoeffner, 2002).

Avoid visual illusions that distort distance and direction.

Visual illusions affect our perception of distances. Vertical lines appear longer than horizontal ones, and straight lines that are interrupted, as in **don't**, appear to be displaced; the diagonal road passing under the highway is actually straight, but it appears to jog under the overpass. This is not a problem in **do**.

[**Don't**]   [**Do**]

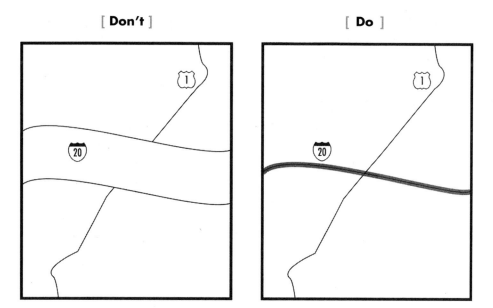

Figure 9.6. Beware of visual illusions (including exaggerated distances of vertical lines) when producing maps. In the map on the left, Route 1 seems to jog as it passes under the Interstate, but it does not.

Do not vary height and width of location markers to specify different variables.

The map in **don't** uses the height of the markers to indicate the population and the width to indicate the mean temperature of each location. But

[**Don't**]   [**Do**]

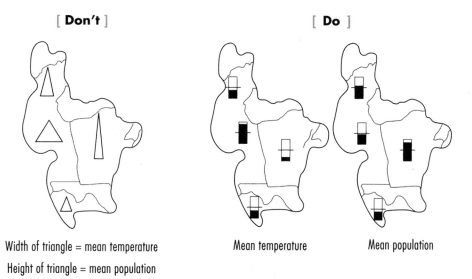

Width of triangle = mean temperature

Height of triangle = mean population

Mean temperature   Mean population

Figure 9.7. If both the width and height of the triangles are varied, we see neither variable very well; it is for better to use two different displays, one for each variable.

the eye sees the area of the markers, not each dimension separately. Do not vary integral dimensions (e.g., height and width, or hue and saturation) to specify values of distinct variables.

Do not vary the area of regions to convey precise quantitative information.

Some years ago it became fashionable to design maps of the United States in which the sizes of the states were varied, as in **don't**, to show relative rates or quantities—per capita beer consumption, number of murders, and so forth; the more per the state, the larger the area. Such maps can convey a rough impression of rank ordering, but fall short if actual amounts or even precise ordering is to be conveyed. We are simply not very good at estimating area and have trouble comparing relative areas of differently shaped regions. A better way to convey amount is to draw a bar or similar symbol at each location; our visual system can compare relative heights or lengths well.

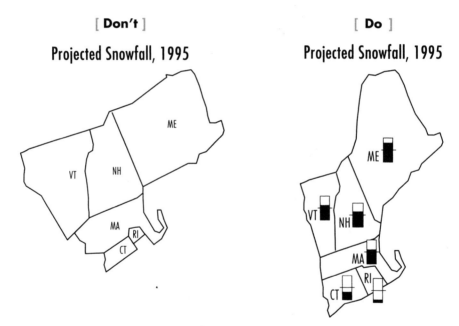

Figure 9.8. It is almost impossible to extract relative amounts from differently shaped areas, even if they are familiar and not distorted. Vary extent along a line or bar instead.

Ensure that regions are identifiable.

Another problem with varying the size of a region to convey information is that the shape may become unrecognizable. If Maine is made huge, its shape may have to be modified to fit it against its neighbors, as happened in the previous **don't** example.

Provide neither more nor less detail than required for the purpose.

The Principle of Relevance is often violated in maps. If you want to illustrate how to get to the fire station from Harvard Square, it would not be helpful to include every road in the region, as in **don't**. Instead, illustrate the roads that are possible routes between the two locations, providing enough landmarks that the traveler won't get lost.

[**Don't**]   [**Do**]

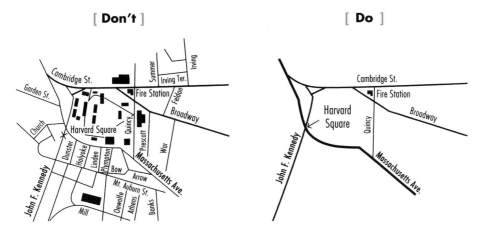

Figure 9.9. A map that includes too much extraneous detail for its purpose is not helpful; to specify the location of the firehouse, only the main routes are necessary.

If distance is important, use grid markings.

Some maps are intended to indicate the relative locations of regions or specific routes, or are intended to provide other qualitative kinds of information. Other maps, however, specify the actual distances between locations. In

[**Don't**]   [**Do**]

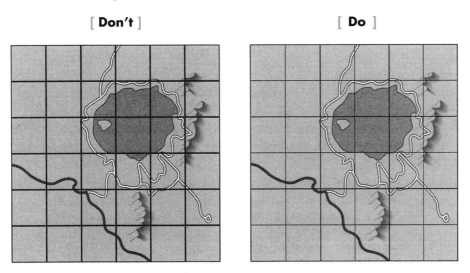

Figure 9.10. If precise distance is not important, there is no reason to

these cases, inner grids are useful and should be drawn following the recommendations presented on pages 182–185: Inner grid lines should be relatively thin and light (this recommendation is violated in **don't**); use more tightly spaced grid lines when greater precision is required; insert heavier grid lines at equal intervals; inner grid lines should pass behind the lines, bars, or other symbols.

The scale should equal one or more grid units.

The scale of a map is a kind of key, pairing a line length with a unit of distance, which the reader uses to estimate the distances between the locations represented on the map. To be effective, the scale should correspond to one or more of the units used to produce the grid lines. If it does not, the reader will have to reorganize the unit line length, mentally dividing it into the corresponding lengths on the grid; this operation is neither efficient nor accurate. If the grid lines are spaced every half inch on the map, the scale should indicate the distance in terms of a familiar multiple of this increment, say, 1 inch. Ideally, the scale should indicate a unit that corresponds to the distance between heavy grid lines, as in **do**.

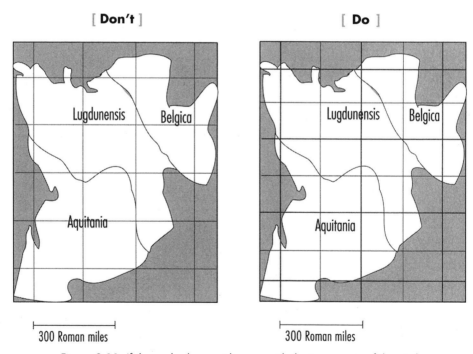

Figure 9.11. If the scale does not line up with the increments of the girds, mental division is necessary to use it; spare the reader this effort by aligning the scale and grid increments.

Make more important routes more salient.

Use the most important routes as the backbone of the map, as in **do**, allowing other routes to be organized by them. This can be accomplished simply by making the more important routes more visually distinctive.

[**Don't**]   [**Do**]

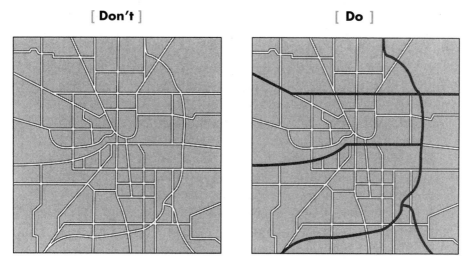

Figure 9.12. If the more salient lines correspond to major roads, the reader is helped to organize the map as well as to find the most efficient routes.

Label important distances directly.

If you want the reader to know the distances between specific pairs of locations, draw lines between them and label them directly (the lines should indicate routes); this recommendation follows from the Principles of Relevance and Informative Changes.

Into the Future

Perhaps the most exciting extensions of the psychological approach to display design take advantage of new technologies. Computer-aided design (CAD) systems allow displays to move in various ways, and motion is another cue human beings use to organize and discriminate among objects. The design of such computerized displays can be made "psychologically optimal" more easily than can charts and graphs because the designer can take full advantage of the dynamic aspects of the computer. Parts of the display can be selected for viewing as needed, with only the relevant information screened at any one time. If you were designing the layout of the electrical system for a new plant, for example, it may be useful from time to time to compare this design with the layout of the plumbing system, but you would not want the two systems always displayed together as you work. Indeed, at one time you may want only to compare one aspect of the plumbing, say, the main sewage trunks, with the layout of the electrical system; at another time, you may want to compare some other aspect of the two systems. In such dynamic computerized displays, the amount of available information can be vast, without overloading our limited-capacity short-term memories or limited processing abilities.

The design of these sorts of dynamic interactive displays would also benefit from the sort of cognitive engineering employed in formulating the recommendations in this book. If one can work out the principles that underlie the "psychological organization" of the object to be displayed—building, car, whatever—one can design the display to take advantage of the way people would break down the object into regions and subsystems. These principles can be used to determine the ways in which specific information is added and deleted from the display.

Such systems must be tailored to the user's needs and capabilities. Tversky, Morrison, and Betrancourt (2002) reviewed the literature on animated graphics and concluded that there is little evidence that animation per se actually aids comprehension or learning. They blame this state of affairs on a failure to respect what they call the "apprehension principle," which states that "the structure and content of the external representation should be readily and accurately perceived and comprehended" (p. 10). This principle appears close to our Principle of Compatibility but also combines elements of many of our principles (which can be viewed as unpacking this rather general admonition).

Essentially, the problem with many animated displays is that they are presented too quickly. As one vivid example of such a limitation, Tversky et al. (2002) note that for centuries people (or, at least, painters) incorrectly believed that the front legs of a galloping horse were extended to the front and the rear legs extended to the rear; the correct positions of the legs was only established after the invention of stop-gap photography. A galloping horse's legs move more quickly than our motion perception mechanisms could track accurately. "To accord with the Principle of Apprehension, animations must be slow and clear enough for observers to perceive movements, changes, and their timing, and to understand the changes in relations between the parts and sequences of events" (p. 12). In the service of this principle, Tversky et al. conclude that animations not only should be slow enough to track, but also should tend to be more schematic than realistic, and should be annotated with symbols (e.g., arrows) that will direct the viewer's attention to crucial changes and relations in the display.

Another way that computers could be brought to bear in future graphics focuses on using computers to adapt to the user's competences and preferences. In order to accomplish this goal, we first would need to understand the nature of the information processing readers use when interpreting displays. In fact, a vigorous interdisciplinary group of researchers has studied how people extract information from information-conveying graphics. Bertin (1983) was perhaps the first to think in detail about the sorts of information processing people perform when they read such graphics. Bertin focused on the kinds of questions one can answer on the basis of graphics. His distinctions, as revised by Wainer (1992), characterize three kinds of questions: (1) those that focus on extracting specific data, (2) those that require abstracting trends in parts of the data, and (3) those that require understanding the overall patterns in the display.

Although Bertin did not take the next step and formulate theories of the perceptual and cognitive processes required to perform these activities, others have since taken up this challenge. For example, Gillan and Lewis (1994) propose that graph reading tasks draw on processes that (1) search for the information-conveying aspects of the display, (2) encode the values conveyed by those marks, (3) perform arithmetic operations on the values, (4) make spatial comparisons among the marks, and (5) produce a response. They also suggest that different types of tasks with different types of graphs evoke different combinations of these processes, which may be in used in different orders. They showed that more time was in fact required to carry out tasks that they expected to require more processing steps. They noted, however, that the generality of their results was unclear because the displays were simplified and the tasks narrowly constrained.

Attali and Goldschmidt (1996) suggest that to understand how people comprehend graphs one must take into account the complexity of the graph itself, the complexity of working with the concepts being displayed, and the processes needed to abstract the required information from the graph. Attali and Goldschmidt provided evidence that these variables do in fact affect the ease of comprehending graphs. Similarly, Friel, Curcio, and Bright (2001) distinguish among three general types of processes. First, some processes simply extract data from a graph; second, other processes interpolate or translate, which leads to integration and interpretation; and third, yet other processes are involved in extrapolating from the depicted data, often for predicting how trends would be extended into the future or across other categories. Curcio (1987) refers to these processes in terms of those that read the data, read between the data, and read beyond the data.

Shah and Hoeffner (2002) offer a different analysis of how people comprehend information graphics. They identify three general types of processes, those involved in (1) encoding the visual array and identifying key components (e.g., a line sloping upward), (2) relating the visual properties to the corresponding conceptual interpretations, and (3) determining the referent of the concepts (what the graph actually means). There is evidence from studies of eye movements to support Shah and Hoeffner's basic ideas (see

Carpenter & Shah, 1998). However, although this scheme breaks down Curcio's second set of processes into two more-finely characterized processes, it does not include the final aspect of processing identified by Curcio and colleagues.

Simkin and Hastie (1987) offer a sophisticated theory of the specific processes that are used to read a graph. This theory specifies four "visual routines" that are "run" on a representation of a graph in order to decode it. The routines allow the reader to find an "anchor point," to scan from that point, to project mentally a reference line (e.g., from one bar to another or from a content element to the Y axis), and to superimpose a mental image of one part of the content over another (e.g., pie wedges, to compare their relative sizes).

Taking a different tack, Lohse (1997) studied the relationship between working memory (WM) capacity and graph reading. WM is a system that holds information in short-term memory and operates on it in some way (Baddeley, 1986). Lohse found that people who have relatively small WM capacity (as assessed with a reading span task) had more difficulty making decisions about a complex display than did people with larger WM capacity. Lohse put it well: "What is effortful for someone with a low WM may not be effortful for someone with a high WM capacity. Therefore, one promising direction for future research is to measure the amount of effort required to extract information from graphic displays" (p. 307).

For example, by respecting individual differences in WM capacity, a display could be designed to adapt to the user, perhaps building up an image a chunk at a time (allowing the user to organize it effectively) for people with relatively low WM. To quote Lohse (1997), "Embedded intelligence in graphics software will remove unnecessary options, customize features and menus to meet individual user needs, make decisions that speed up the graph-creation process, and ensure that certain rules for effective graphic design are followed" (p. 307).

In addition, there is no reason why a computer could not present the same data in several different ways—either alone or in combination. This would be desirable because Carswell, Bates, Pregliasco, Lonon, and Urban (1998) found that people not only differ in their preferences for using graphs, but also tend to use different strategies when interpreting them (see also Carswell & Ramzy, 1997). Carswell et al. (1998) suggest that these differences may derive in large part from differences in the goals users bring to the task of interpreting a graph. If so, then depending on one's goal, one may prefer a different type of display.

Another direction for future research is developing good computer simulation models of graph reading and comprehension. This may be the only way in which we will be able to produce precise advice about how to display specific data for a specific purpose. As Meyer (2000) notes, it currently is impossible to offer hard-and-fast rules for how to make graphs, or even when to use them. The problem is that a myriad of variables are important, and these variables interact in complex ways (e.g., Coll, Coll, & Thakur, 1994; Meyer, Shinar, & Leiser, 1997). For all practical purposes, it is impossible to conduct definitive experiments; simply too many variations in stimulus conditions are possible. Rather, Meyer (2000) suggests that models are the way to go. If we have good computer simulations of how people read graphs, we need not actually conduct the experiments—we can just run the model, which will tell us how best to use graphics in any specific situation.

However, this goal will not be easy to attain. In my view, at present we do not know enough even to begin to build such models; the attempts so far have not been terribly encouraging (e.g., see Meyer, 2000; Trickett, Ratwani & Trafton, 2005). Nevertheless, the effort of constructing such models is worthwhile, if only because it may help to organize the laboratory research effort (e.g., Carpenter & Shah, 1998). Instead of randomly combining variables, by considering ways in which models could be developed, we will be directed to examine specific aspects of graph perception and comprehension.

In closing, I wish to leave the reader with an optimistic observation: The empirical study of visual communication is flourishing, and the fruits of this research are producing solid guidelines for developing information-graphics for specific purposes. As research on perception and cognition (both general and specifically focused on information graphics) progresses, designers will be increasingly well placed to take advantage of readers' perceptual and cognitive strengths and avoid falling prey to their weaknesses. And hence readers will be increasingly often treated to displays that illuminate and inform, that are coherent, compelling, and crystal clear.

Appendix 1

Elementary Statistics
for Graphs

R. S. Rosenberg and S. M. Kosslyn

This appendix provides a brief review of statistical concepts that are used in this book.

Variables

A *variable* is a measurable characteristic of a substance, quantity, or entity. The simplest form of a quantitative relationship measures one variable against another: heat in degrees against time, sales in dollars against season, tax revenue against each of several cities. Looking at these pairings, we can see that the values of one variable will change in relation to the values of the other: How hot the soup is depends (up to a point) on how long it's been on the burner; the sales depend on the time of year; the tax revenue depends on whether the city is Los Angeles or Toledo. Therefore, one of the variables— the quantity whose changes we want to observe—is called the *dependent variable*; the other is the *independent variable*. In a line graph, for example, the dependent variable is graphed on the Y axis and the independent variable is graphed on the X axis.

But consider again the examples. "Time," "temperature," and "revenue" are not quite like "city." Time, temperature, and revenue are *continuous* variables; the values are numbers that change from one into the other. You can perform mathematical operations on these numbers, adding or subtracting two values, multiplying them, and so forth. In contrast, "city" is a *categorical* variable. It does vary—it can be either Tokyo or Toledo; but it does not vary continuously, it does not change from one into the other, and you can't perform mathematical operations on these values. Commonly used categorical variables are gender, location, team, and political party.

Scales

Variables are measured along several different types of scales; traditionally, there are four types of scales, each with its own properties and arithmetic restrictions. A *nominal scale* (also called a categorical scale) presents values of categorical variables: names of the individuals, groups, or categories for which data will be shown. There may be numbers assigned to such a scale—the uniform numbers of the players on a team, or numbers of television channels—but these numbers function as names and cannot be manipulated arithmetically. This kind of scale may be arranged with a sense of number (e.g., you may choose to arrange the cities for which you are plotting data from smallest to largest), but the members of the scale are discrete, and the scale is in no sense continuous.

An *ordinal scale* arranges data by rank, ordering it into first, second, third, and so forth. Candidates for political office could be ranked by their standings in opinion polls; runners, by their finish places. Rank is all that an ordinal scale shows: You cannot assume that the first-place candidate or runner is as far ahead of the second as the second is of the third—that is, you cannot assume equal intervals between the ranks. Only the ordering of the entities is important.

To know and compare intervals, you need an *interval* scale or its close cousin, a *ratio* scale. Both measure equal distances along a continuous scale, and the distinction between them is subtle. On a ratio scale, there is set starting point to which the quantities are measured, an absolute zero, and so meaningful ratios can be computed from the intervals: Two weeks is twice as long as one week, ten pesos is twice as much as five pesos, and—on the Kelvin scale, which begins at absolute zero, the "starting point" of heat—forty degrees is twice as hot as twenty degrees. The last example pinpoints the difference between ratio and interval scales: When temperature is measured on the centigrade or Fahrenheit scale, neither of which is calibrated from absolute zero, forty degrees is *not* twice as hot as twenty degrees, merely twenty degrees hotter. Data on a ratio scale permit more arithmetical and statistical operations than do data on the other three scales. For purposes of graphing data, however, the distinction between ratio and interval scales is usually not important, and indeed, it has not been made in the discussions in this book.

Expressing Values

Depending on how data are construed, different information will be revealed. It is often useful to know the *central tendency* of a set of numbers, the locus of the most typical values, or *scores*, for a given sample or group.

There are three common ways to express central tendency. The most frequently used is the *mean*, or arithmetic average, which is obtained by adding all the scores in a group and dividing the sum by the total number of scores: using the values in table A.1, $(7 + 1 + 2 + 8 + 1 + 4 + 5)/7 = 4.0$. The *median* is the value at the midpoint of a series of scores—half the total number of scores fall above that value, and half fall below. Arranging the numbers in table A.1 as 1 1 2 4 5 7 8, we see that the median is 4. The mode, the value that occurs most frequently in the sample, may fall at the high end or the low end of the series, or anywhere along it. In our example the mode is 1. In most cases, however, the mode usually falls on or near the other two measures. The mean, median, and mode can be computed on ordinal, interval, and ratio scales; the mode can also be computed for nominal scales, but it must be stated slightly differently—the "most frequent value" on a nominal scale might be read as "most [of the group] are women" or "most [of the group] are third-graders." The mean is the most sensitive to extreme values, the mode the least sensitive. In general, a median is more appropriate than a mean if the data are not distributed along the familiar bell-shaped curve (called a *normal* distribution), on which most values fall in the middle range and the extremes taper off in symmetrical tails.

If scores are presented directly, not expressed as measures of central tendency, they are raw data. They may be transformed in various ways to provide more information, indicating where they fall relative to one another. A *percentile* rank of a given score specifies the percentage of cases scoring at or below that score. Converting or transforming a score to a percentile rank immediately conveys where that score falls compared to the rest of the group or data set. Quartiles divide the group into fourths—the 25th, 50th, 75th, and 100th percentiles—so a score at the first quartile (or the 25th percentile) signifies that twenty-five percent of the sample fall at or below that score.

Table A.1. Daily Production of Widgets
at Seven Factories

Factory	Number of Widgets	Squared Deviations
A	7	$(7-4)^2 = 3.0^2 = 90$
B	1	$(1-4)^2 = -3.0^2 = 9.0$
C	2	$(2-4)^2 = -2.0 = 4.0$
D	8	$(8-4)^2 = 4.0^2 = 16.0$
E	1	$(1-4)^2 = -3.0^2 = 9.0$
F	4	$(4-4)^2 = 0.0^2 = 0$
G	5	$(5-4)^2 = 1.0^2 = 1.0$
Cases = 7	Total = 28	48.0 = Sum of squares (SS)

Mean: Total number of widgets/number of cases =28/7 = 4.0 widgets
Median: Value representing 50th percentile = 4 widgets
Mode: Most frequently occurring value = 1 widget
Range: Largest value – smallest value = 8 – 1 = 7 widgets
Variance (corrected): Sum of squares/(number of cases – 1) = 48.0/6 = 8.00
Standard deviation (corrected): Square root of variance: square root of 8 = 2.83
Standard error of the mean (corrected): Standard deviation/square root of (number of cases –1) = $2.83/\sqrt{6}$ = 2.83/2.45 = 1.16

Similarly, deciles divide the group into tenths: A score at the third decile (30th percentile) signifies that thirty percent of the sample falls at or below that score.

Another method of transforming scales is by using logarithms. This is a useful technique if the scores have extreme values and important variations occur among small values. The logarithm, usually based on 10 or 2, indicates the amount that the base value would have to be raised to produce a given number. If base 10 is used, the log of 10 is 1, the log of 100 is 2, and so forth. The amount that 10 would be raised is specified by the exponent. If 10 is raised by a factor of 2, the exponent is 2: $10^2 = 100$. Ten is a convenient base because the exponent indicates how many zeros follow the digit 1. Fractions of an exponent can be used, so all numerical values can be expressed this way. If logarithms are plotted, the equivalent distance on a scale represents increasing amounts as one ascends the axis. This is of interest in creating graphs because by using a log scale, equivalent distances on the axis represent progressively increasing amounts: The first interval represents 10 units, the next 10^2 or 100 units, the next 10^3 or 1,000 units. Differences among large numbers are compressed, because the same distance on the scale stands for increasingly large increments as numbers get larger. As a consequence, more of the Y axis is available to display variations among small values than would be available if the untransformed values were plotted.

Measures of Variability

Variability conveys information about the spread or scatter of the scores. The simplest measure of variability is the *range*, which is obtained by subtracting the smallest score or value in the set from the largest score or value. Like the mean, the range is sensitive to extreme scores and does not give information about how the variability is distributed over the scores. Two factories may have the same range in annual income of managers, but most of the managers at factory A are at the lower end of the pay scale, whereas the salaries of managers at factory B are scattered equally throughout the pay scale. Expressing the range would not capture this difference.

You can also express variability by presenting information about the range of some proportion of the data. For example, if medians are graphed, you could use "I"-shaped brackets to plot the values of scores that are twenty percent of the way toward the extreme values. This practice will be most effective, however, if readers have some idea of the stability of such measures.

Perhaps the most common method of expressing variability is the *standard deviation*, which describes the average variability for a group or set. It is based on the average of the deviations of each score around the mean, which are called the deviation scores. More specifically, because the sum of the difference of each score from the mean of those scores will always equal zero, the standard deviation is computed by squaring each deviation score (i.e., the difference between each score and the mean) and then summing these squared values; this is called the sum of squares (in most statistics textbooks abbreviated SS). That sum divided by the number of deviation scores used to obtain it produces the *variance*. However, to be a useful measure of variability that can be associated with a measure of central tendency, we must take the square root of the variance (remember, we squared the deviation scores at the beginning so we must "undo" the square); this square root is the standard deviation.

There are two standard ways to estimate the variance. The one just described is sometimes used in summaries of central tendency and variability. But if you want the reader to be able to decide whether the difference between two means is statistically significant (reflects an actual difference between the measured entities, rather than an accident of how the measurements were obtained), the variance must be computed by dividing not the total number of cases that went into it but that number minus 1. (This is a truer measure because the variance from a sample is an underestimate of that from the entire population; subtracting 1 from the total number of cases is a correction factor, producing an unbiased estimate of the variance.) The square root of this variance can be taken in turn to produce a standard deviation. This measure is appropriate to plot.

Another measure of variability is the *standard error of the mean*, which reflects the variability of means of your sample. Imagine taking a set of 10 apples at random from a supermarket bin, weighing them, and then taking the mean. Then take another set of 10 apples, weigh them, and take their mean. The two means would not be exactly the same. If you took a third sample, its mean would probably be a little different from the other two, and so on. You can estimate the variability of these means simply by taking the standard deviation of one sample (one set of 10 apples) and dividing it by the square root of the number of cases (10, in the example). You can then plot one standard error of the mean above and below a point on a graph.

It is a good idea to plot the standard error of the mean on graphs because it is a comment on the data you are presenting. The reader sees means, indicated by points or dots, and wants to know how stable they are—that is, how seriously to take them. The standard error of the mean indicates the range in which one could expect to find the mean if repeated samples were taken.

Best-Fitting Lines

A scatterplot is often most useful if the trend in the data is highlighted by a line fitting through the cloud of points. Depending on the relationship between the variables being plotted, this may be a straight line or one of several curves. The line is, in a sense, the ideal representation of the trend: If no data point deviated from the trend, all the points would lie on the line.

But they don't. The problem is to lay the line along a path from which the data points deviate the least. The most common way of doing this is by the *method of least squares*. The vertical deviation of each point, whether above or below the line, is squared (to eliminate positive and negative values that would cancel each other out), and the squares are summed. The line is in fact an ideal representation of the trend—that it, it is best fitted—when the sum of the squares is the smallest possible. Needless to say, a trial-and-error approach to the method of least squares is a formidable prospect, and in practice the line is placed by algebraic formula, and a calculator with statistical functions (or a computer with a suitable graphs program) is of inestimable value.

A related concept is the *correlation*, usually expressed by r, which provides a measure by which to judge how well measurements on one scale are predicted by measurements on the other. If $r = 1.0$, the points would fall perfectly on a best-fitting line; if $r = 0$, the relationship would be random. Again, this value is found by using algebra. The correlation value squared tells you what percentage of the variability in the values on one scale is accounted for by the variability in the values on the other scale. For example, if $r = 0.80$, then $r^2 = 0.64$, and sixty-four percent of the variation in the values of the data points can be explained by the relationship between the two variables. Correlations are symmetrical; values of either variable can be used to predict values of the other.

Lying With Statistics

Graphs can be misleading in part because misleading numbers are presented (as was shown brilliantly by Darrell Huff in his 1954 book, *How to Lie With Statistics*). Different statistics or comparisons represent different facets of the data. Suppose you are the public relations expert for a political candidate. Part of your job is to issue press releases announcing the latest standing of your candidate in the polls. There are many ways you could do this. You could report the actual number (absolute value), or the percentage of people who favored your candidate. You could compare this number (or percentage) with many other numbers, depending upon the point you wanted to illustrate; for example, you could compare it to your opponent's standing, your candidate's standing last month, the projections from last month, the projection of how many votes (or what percentage) are needed to win, or your candidate's all-time low or all-time high standing. You could report not the actual number or percentage of those people favoring your candidate, but the percentage increase. This would be especially useful if your candidate, who has been trailing in the polls, increased from five to ten percent in share of the projected vote: You can state that your candidate's popularity has doubled. You could report your candidate's mean standing in polls, the median, or the mode. Each choice illustrates a different point.

As you can see, even before you start graphing the data, the statistics that you choose to represent the data can create a bias in how they will look and the message the reader will receive. Be a skeptical reader and ask yourself this question: If the data were presented using different statistics, or compared to a different referent, would the message be different?

Appendix 2

Analyzing Graphics Programs

New or revised graphics programs for personal computers are released practically every other month, and it would be silly to try to provide a comprehensive review of all graphics programs available. It would be equally silly to ignore them and not formulate a means of review. The checklist below is intended to provide a structured way to evaluate graphics programs and choose one suitable for your particular needs. It would have been possible (but not useful—recall the Principle of Relevance!) to develop a question for each recommendation offered in this book. Instead, the questions presented are meant to identify how well a program produces the most frequently used display formats, and to isolate aspects and features that may be of particular importance to you. The checklist is designed to help you compare the relative virtues and drawbacks of a number of programs; it is not meant to make up your mind for you, but to focus your attention on what's important to you in a graphics program.

To use the checklist, score each *yes* answer with 1 point if the feature in question is one you would use sometimes, with 2 points if that feature is very important to you. Similarly, subtract 1 or 2 points as appropriate for each *no* answer. If the feature is of no concern to you, no points need be added or subtracted; alternatively, if the presence of a feature is a strong selling point with you, you may wish to award it more than 2 points. Comparison of the total scores awarded to different programs will tell you how useful a given program is for your purposes.

The checklist is in four parts. The first focuses on how easy it is to use a program, without considering the quality of the displays it produces. The second, third, and fourth parts deal with specific aspects of a program that make its products more or less effective. These last three parts of the checklist will bring to light most of the important features of a program.

Ease of Use

Talk to someone who uses the program you are evaluating. Many of these questions are best answered by the experience of someone who has passed through the initial stages of learning a program and has become adept at its use.

1. Does the program produce the display formats you use?
2. Does the program allow the generation of novel types of displays?
3. Can usable graphs be produced within 15 minutes of entering data?
4. Does the display on the screen look the same as what is printed out? [The WYSIWYG ("what you see is what you get") approach is popular in the Macintosh world but not in the UNIX world; programs for PC machines vary widely in this respect. WYSIWYG is strongly recommended for display design because you will sometimes have to make several successive drafts; if this can be done on the screen, so much the better.]
5. Do you see the graph as it is being set up, so that you can alter its design as soon as a flaw appears?
6. Does the program require less than an average of five minutes to revise any aspect of a simple line graph?
7. Does the program save most of the information used to plot a graph so you do not have to reenter it in order to plot data in a new way?
8. Is the program integrated with an effective, useful, spreadsheet program?
9. Can data be entered in more than one way? [After you are familiar with the program, you probably will want to enter data in a batch mode (as a separate file, rather than by retyping it a part at a time).]
10. Does the program have many ways of entering commands? [This sort of flexibility should not be underestimated. Although interactive modes are helpful in some circumstances, especially when you are first learning, using them is often slower than specifying lots of information in a single batch at the outset.]
11. Does the program make it easy to plot subsets of data?

12. Can you easily alter which independent variable is on the X axis and which is the parameter (e.g., plotted as separate lines; see chapter 1)?
13. Can you easily alter what type of content—bars, lines, or points—is used?
14. Does the program have a useful Help function? [A useful Help function is one that can be accessed without disrupting what you are doing, is organized so that one can easily find the information needed, and does not assume that you know too much.] Is the program so well designed that you do not need to use the Help function?
15. Is the program so well designed that you do not need to read more than a few pages of instructions?
16. Is the program written by a reputable software house, which is likely to support it over the long term and release periodic upgrades?

Creating the Framework

Graphs in L- or T-shaped frameworks

1. Does the program automatically scale the range of values along the Y axis?
2. Can you override a function that automatically scales the range of values along the Y axis?
3. Can you control the origin of the Y axis?
4. Are tick marks placed at regular intervals (e.g., multiples of 10), instead of at locations of plotted values?
5. Are tick mark labels always round numbers?
6. Is every fifth or tenth tick mark drawn more heavily?
7. Can an inner grid be added or omitted?
8. Is the inner grid drawn with finer lines than used for the outer framework?
9. Is it possible to add a right Y axis to an L-shaped framework?

Pie graphs

1. Can you control which slice or slices are exploded?
2. Can pies can be drawn using two dimensions, or must they be drawn in three?
3. Is genuine projective 3-D used, not "pseudo 3-D" (which simply portrays parts as seen from different points of view)?

Creating the Content

1. Can you control the hue, saturation, and intensity of colors used for content elements?

2. Does the program automatically use discriminable colors for different content elements?
3. Does the program automatically vary intensities of colors?
4. Can the program automatically use colors that have about the same salience?

Bar graphs

1. Are bars automatically drawn with clearly discriminable internal hatching?
2. When three-dimensional bars are drawn, are they placed in a three-dimensional framework?
3. Can you adjust the spacing of bars?
4. Can you adjust the width of the bars?
5. Can you easily change the order of the bars?

Line graphs and scatterplots

1. Are the leftmost and rightmost dots always plotted within the framework?
2. Are lines automatically drawn in clearly different patterns?
3. Are the dots in line graphs or scatterplots automatically drawn using clearly different symbols when appropriate?
4. Can you control which symbols are used to plot lines and dots?

Step, pie, layer, stacked-bar, and divided-bar graphs

1. Are different regions (wedges, slices, steps, etc.) automatically drawn so that they are visually distinct?
2. Can you control how regions are filled with shading or hatching?
3. Can you control how segments are ordered?

Creating the Labels

A good program will place labels correctly, as described in chapter 3; it will also allow the user to adjust labels to prevent clutter, ensure detectability, and so on.

1. Does the program allow you to move labels?
2. Can the size and other typographic properties of labels be adjusted?
3. Is the title automatically drawn using letters that are typographically distinct from the other labels (but not all uppercase letters)?
4. Are the smallest labels clearly readable when printed out?
5. Are discriminable colors used to associate labels with content elements?

6. Can you control the salience of colors used to group labels and content elements?

Graphs in L- or T-shaped frameworks

1. Are axis labels centered next to the relevant axis?
2. Can you add spaces where you want them between labels and content elements?
3. Are even long labels clearly associated with the correct tick mark on the X axis?
4. Are labels placed closest to the correct content element, and do they all appear in the same region of the display?
5. Can you choose when to use a key?
6. Are labels automatically placed so that they do not cut across an axis or content element?

Pie graphs

1. Are labels automatically placed consistently for each wedge, or are some inside and some outside the wedges?
2. Can you control the locations of labels?
3. Can you control whether or not a key is used?

Creating Multiple Panels

1. Can you choose which data to plot in multiple panels?
2. Can you easily arrange multiple panels on a page?
3. Are corresponding segments automatically ordered the same way in multiple displays?
4. Can you adjust the scales of the individual panels?

Appendix 3

Summary of Psychological Principles

The Eight-Fold Way hinges on applying the following principles to the design of visual displays.

Principle of Relevance
Communication is most effective when neither too much nor too little information is presented.

Readers expect to see all and only the information that is relevant to the purpose of the display. Presenting too little information will puzzle the readers, and presenting too much will overwhelm them with needless detail. Thus, before beginning to design or produce a display, you need to be clear on what message you want to convey; only after you have made this decision can you decide what information to include.

Principle of Appropriate Knowledge
Communication requires prior knowledge of relevant concepts, jargon, and symbols.

You will communicate effectively only if you make use of what the readers already know. You must "know your audience." An effective display must be pitched at the right level for the readers you wish to reach. If your graph is for fellow rocket scientists, you can (and should) use those specialized concepts, display formats, and those esoteric symbols; if it's for *USA Today*, you should use common formats and you will need to explain yourself, and probably stick to the central idea and forgo many of the details. In addition, a good display presents new information—but not too new. Readers can interpret a display only if it builds on appropriate information that they have already stored in memory, such as information about the way the framework represents variables and how the content relates to specific parts of the framework.

Principle of Salience
Attention is drawn to large perceptible differences.

The most visually striking aspects of a display will draw attention to them, and hence they should signal the most important information. All visual properties are relative, and thus what counts as "visually striking" depends on the properties of the display as a whole.

Principle of Discriminability
Two properties must differ by a large enough proportion or they will not be distinguished.

Detectability is special case of this principle: Marks must be large or heavy enough to be noticed (i.e., distinguished from the background).

Principle of Perceptual Organization
People automatically group elements into units, which they then attend to and remember.

Our visual systems are not like cameras, which record what they are pointed at in a relatively veridical way. Instead, we actively organize and interpret what we see, and the display designer must respect key characteristics of such processing or the mental representation will not preserve the information the designer intended to convey. What follows are aspects of processing that a designer should be aware of, in order to ensure that the display produces the appropriate mental representation.

Input Channels. A reader will "automatically" pay attention to patterns that are registered by the same "input channel," and effort is required to attend to individual components of such patterns. The visual system responds to variations at multiple levels of scale (levels of acuity) and in orientations (orientation sensitivity). First, *levels of acuity*: The visual system acts as if it processes the outputs from separate lenses, which differ in the scope and detail of what they register. Each "channel" is sensitive to changes in regular light/dark alterations within about a 2-to-1 ratio. Second, *orientation sensitivity*: Each "orientation channel" registers orientations within a range of about thirty degrees. Hence, to be immediately distinguished, orientations must differ by at least this amount.

Three-Dimensional Interpretation. If at all possible, we interpret a pattern as representing a three-dimensional object. Thus, three dimensions can be depicted on a flat surface, using devices such as foreshortening and texture gradients; however, the effect is not perfect, and quantitative differences in depth are not accurately portrayed.

Integrated Versus Separated Dimensions. Integral dimensions (e.g., height and width, or hue and saturation) are "automatically" integrated by the visual system; hence, values on one dimension affect how one sees values on the other. In contrast, separable dimensions are processed individually, without affecting each other. A reader can attend to individual integral dimensions when another varies, but only with considerable effort and not very effectively.

Grouping Laws. The visual system also automatically groups input into psychological units according to a set of "grouping laws". Marks that are nearby (*proximity*), arranged in a continuous pathway (*good continuation*), alike in shape, color, or other visual characteristics (*similarity*), moving in the same way or direction (*common fate*), or arranged in regular patterns (*good form*) are grouped into single perceptual units.

Principle of Compatibility
A message is easiest to understand if its form is compatible with its meaning.

The appearance of a pattern should be compatible with what it symbolizes. The injunction *not* to judge a book by its cover is an attempt to fight our natural tendency to do just that; we take appearance as a clue to the reality. This principle has the following aspects.

Surface–Content Correspondence. What the reader sees is what the reader gets. The Stroop experiment (p. 15) showed dramatically that if the physical appearance and interpretation conflict (in this case, the color of ink and meaning of the word written in it), it is more difficult to process. The mind attempts to fit all the information presented to it into a single coherent framework and expends effort to do so.

More Is More. A greater amount of a perceptible quality, such as height or area, should represent a larger quantity.

Perceptual Distortion. Some visual dimensions are systematically distorted; notably, area, intensity, and volume are progressively underestimated as they increase. In contrast, the lengths of lines at the same orientation are registered relatively accurately, although vertical lines appear longer than horizontal ones of the same length.

Spatial Imprecision. In addition, some distortion arises because of the inherent imprecision of the brain. Notably, shapes and locations are not always precisely conjoined during perception. Imprecision in judging spatial relations apparently occurs because different brain systems are used to register object properties (e.g., shape, color, and texture) and spatial properties (e.g., location, size, and orientation).

Cultural Convention. The Principle of Compatibility applies even when a meaning arises from common *cultural conventions*; for example, red means "stop" and green means "go," in Western cultures, and thus printing the word "stop" in green and "go" in red would violate this principle.

Principle of Informative Changes
People expect changes in properties to carry information.

The reader will interpret any change in the appearance of a display (changing the color or texture, adding or deleting lines, etc.) as conveying information. By the same token, the reader will expect any piece of information that should be conveyed in a display to be indicated by a visible change in the display (e.g., a mark should clearly indicate where a scale has been truncated; a change from current to projected data should be clearly indicated).

Principle of Capacity Limitations
People have a limited capacity to retain and to process information and will not understand a message if too much information must be retained or processed.

This principle has two major aspects.

Short-Term Memory Limits. A display will not be understood if it requires readers to hold in mind too much information at the same time. We can hold in mind only about four units at the same time.

Processing Limits. A display will not be understood if it requires readers to expend too much effort to process it. Searching a display requires effort, and mental transformation (e.g., visualizing a line connecting the tops of bars), addition, subtraction, comparison, or averaging operations also require readers to expend effort.

References

Aldrich, F. K., & Parkin, A. J. (1987). Tangible line graphs: An experimental investigation of three formats using capsule paper. *Human Factors, 29*, 301–309.

Allen, R. C., & Rubin, M. L. (1981). Chromostereopsis. *Survey of Ophthalmology, 26*, 22–27.

Anderson, J. R. (Ed.). (1981). *Cognitive skills and their acquisition.* Hillsdale, NJ: Erlbaum.

Anderson, J. R. (2004). *Cognitive psychology and its implications, sixth edition.* New York: Worth.

Arditi, A., Knoblauch, K., & Grunwald, I. (1990). Reading with fixed and variable pitch. *Journal of the Optical Society of America A, 7*, 2011–2015.

Aretz, A. J., & Calhoun, G. L. (1982). Computer generated pictorial stores management displays for fighter aircraft. In *Proceedings of the Human Factors Society 26th Annual Meeting* (pp. 455–459). Santa Monica, CA: Human Factors Society.

Attali, Y., & Goldschmidt, C. (1996). The effects of component variables on performance in graph comprehension tests. *Journal of Educational Measurement, 33*, 93–105.

Baddeley, A. D. (1986). *Working memory.* Oxford: Oxford University Press.

Baird, J. C. (1970). *Psychophysical analysis of visual space.* New York: Pergamon Press.

Barfield, W., & Robless, R. (1989). The effects of two- and three-dimensional graphics on the problem-solving performance of experienced and novice decision makers. *Behavior & Information Technology, 8*, 369–385.

Barnett, B. J., & Wickens, C. D. (1988). Display proximity in multicue information integration: The benefits of boxes. *Human Factors, 30*, 15–24.

Beeby, A. W., & Taylor, H. P. J. (1973). How well can we use graphs? *Communication of Scientific & Technical Information, 17*, 7–11.

Beringer, D. B. (1987). Peripheral integrated status displays. *Displays, 1,* 33–36.

Bernard, M. L., Lida, B., Riley, S., Hackler, T., & Janzen, K. (2002). *A comparison of popular online fonts: Which size and type is best?* Owner: Usability News 4.1. Created: 2002. Retrieved: 12 December 2005. http://psychology.wichita.edu/surl/usabilitynews/41/onlinetext.htm

Bernard, M., Mills, M., Peterson, M., & Storrer, K. (2001). *A comparison of popular online fonts: Which is best and when?* Owner: Usability News 3.2 Created 2001. Retrieved 12 December 2005. http://psychology.wichita.edu/surl/usabilitynews/3S/font.htm

Bertin, J. (1983). *Semiology of graphs* (W. J. Berg, Trans.). Madison, WI: University of Wisconsin Press.

Bickel, P. J., Hammel, E. A., & O'Connell, J. W. (1975). Sex bias in graduate admissions: Data from Berkeley. *Science, 187,* 398–404.

Birren, F. (1961). *Color psychology and color therapy.* New Hyde Park, NY: University Books.

Bloomer, C. M. (1990). *Principles of visual perception* (2nd ed.). New York: Design Press.

Boyarski, D., Neuwirth, C., Forlizzi, J., & Regli, S. H. (1998). A study of fonts designed for screen display. In *Proceedings of CHI'98* (pp. 87–94). Los Angeles, CA: ACM Press.

Boynton, R. M., Fargo, L., Olson, C. X., & Smallman, H. S. (1989). Category effects in color memory. *Color Research & Application, 14,* 229–234.

Brinton, W. C. (1916). *Graphic methods of presenting facts.* New York: Engineering Magazine.

Brockmann, R. J. (1991). The unbearable distraction of color. *IEEE Transactions on Professional Communication, 34,* 153–159.

Brown, R. L. (1985). Methods for graphic representation of simulated data. *Ergonomics, 28,* 1439–1454.

Bryant, P. E., & Somerville, S. C. (1986). The spatial demands of graphs. *British Journal of Psychology, 77,* 187–197.

Carpenter, P. A., & Shah, P. (1998). A model of the perceptual and conceptual processes in graph comprehension. *Journal of Experimental Psychology: Applied, 4,* 75–100.

Carswell, C. M. (1991). Boutique data graphics: Perspectives on using depth to embellish data displays. In *Proceedings of the Human Factors Society 35th Annual Meeting* (pp. 1532–1536). Santa Monica, CA: Human Factors and Ergonomics Society.

Carswell, C. M. (1992a). Choosing specifiers: An evaluation of the basic tasks model of graphical perception. *Human Factors, 34,* 535–554.

Carswell, C. M. (1992b). Reading graphs: Interactions of processing requirements and stimulus structure. In B. Burns (Ed.), *Percepts, concepts and categories* (pp. 605–645). Amsterdam: Elsevier.

Carswell, C. M., Bates, J. R., Pregliasco, N. R., Lonon, A., & Urban, J. (1998). Finding graphs useful: Linking preference to performance for one cognitive tool. *International Journal of Cognitive Technology, 3,* 4–18.

Carswell, C. M., Frankenberger, S., & Bernhard, D. (1991). Graphing in depth: Perspectives on the use of three-dimensional graphs to represent lower-dimensional data. *Behaviour & Information Technology, 10,* 459–474.

Carswell, C. M., & Ramzy, C. (1997). Graphing small data sets: Should we bother? *Behaviour & Information Technology, 16,* 61–71.

Carswell, C. M., & Wickens, C. D. (1987a). Information integration and the object display: An interaction of task demands and display superiority. *Ergonomics, 30,* 511–528.

Carswell, C. M., & Wickens, C. D. (1987b). Objections to objects: Limitations of human performance in the use of iconic graphics. In L. S. Mark, J. S. Warm, & R. L. Huston (Eds.), *Ergonomics and human factors: Recent research* (pp. 253–260). New York: Springer-Verlag.

Carswell, C. M., & Wickens, C. D. (1988). Comparative graphics: History and applications of perceptual integrality theory and the proximity compatibility hypothesis. Technical Report. Urbana-Champaign: Aviation Research Lab, Institute of Aviation, University of Illinois.

Carswell, C. M., & Wickens, C. D. (1990). The perceptual interaction of graphic attributes: Configurality, stimulus homogeneity, and object integration. *Perception & Psychophysics, 47*, 157–168.

Carter, L. F. (1947). An experiment on the design of tables and graphs used for presenting numerical data. *Journal of Applied Psychology, 31*, 640–650.

Casali, J. G., & Gaylin, K. B. (1988). Selected graph design variables in four interpretation tasks: A microcomputer-based pilot study. *Behaviour & Information Technology, 7*, 31–49.

Casey, E. J., & Wickens, C. D. (1986). *Visual display representation of multidimensional systems.* Urbana-Champaign, IL: Cognitive Psychophysiology Laboratory, University of Illinois.

Casner, S., & Larkin, J. H. (1989). Cognitive efficiency considerations for good graphic design. *Proceedings of the Cognitive Science Society* (pp. 275–282). Hillsdale, NJ: Erlbaum Associates.

Cavanagh, P., Anstis, S., & Mather, G. (1984). Screening for color blindness using optokinetic nystagmus. *Investigative Ophthalmology & Visual Science, 25*, 463–466.

Chambers, J. M., Cleveland, W. S., Kleiner, B., & Tukey, P. A. (1983). *Graphical methods for data analysis.* Belmont, CA: Wadsworth.

Chen, L. (1982). Topological structure in visual perception. *Science, 218*, 699–700.

Chernoff, H. (1973). The use of faces to represent points in k-dimensional space graphically. *Journal of the American Statistical Association, 68*, 361–368.

Chernoff, H., & Rizvi, H. M. (1975). Effect on classification error of random permutations of features in representing multivariate data by faces. *Journal of the American Statistical Association, 70*, 548–554.

Christner, C. A., & Ray, H. W. (1961). An evaluation of the effect of selected combinations of target and background coding on map-reading performance—experiment V. *Human Factors, 3*, 131–146.

Clark, H. H., & Clark, E. V. (1977). *Psychology and language: An introduction to psycholinguistics.* New York: Harcourt, Brace, Jovanovich.

Clement, D. E. (1978). Perceptual structure and selection. In E. C. Carterette & M. P. Friedman (Eds.), *Handbook of perception* (Vol. 9, pp. 49–84). New York: Academic Press.

Cleveland, W. S. (1985). *The elements of graphing data.* Monterey, CA: Wadsworth.

Cleveland, W. S. (1987). Research in statistical graphics. *Journal of the American Statistical Association, 82*, 419–423.

Cleveland, W. S., Diaconis, P., & McGill, R. (1982). Variables on scatterplots look more highly correlated when the scales are increased. *Science, 216*, 1138–1141.

Cleveland, W. S., Harris, C. S., & McGill, R. (1982). Judgments of circle sizes on statistical maps. *Journal of the American Statistical Association, 77*, 541–547.

Cleveland, W. S., & McGill, R. (1984a). Graphical perception: Theory, experimentation, and application to the development of graphical methods. *Journal of the American Statistical Association, 79*, 531–554.

Cleveland, W. S., & McGill, R. (1984b). The many faces of a scatterplot. *Journal of the American Statistical Association, 79*, 807–822.

Cleveland, W. S., & McGill, R. (1985). Graphical perception and graphical methods for analyzing scientific data. *Science, 229*, 828–833.

Cleveland, W. S., & McGill, R. (1986). An experiment in graphical perception. *International Journal of Man-Machine Studies, 25*, 491–500.

Cleveland, W. S., & McGill, R. (1987). Graphical perception: The visual decoding of quantitative information on graphical displays of data. *Journal of the Royal Statistical Society, 150* (Series A, Part 3), 192–229.

Coffey, J. L. (1961). A comparison of vertical and horizontal arrangements of alphanumeric material—experiment 1. *Human Factors, 3*, 93–98.

Coll, R. A., Coll, J. H., & Thakur, G. (1994). Graphs and tables: a four-factor experiment. *Communications of the ACM, 37*, 77–86.

Condover, D. W., & Kraft, C. L. (1954). *The use of color in coding displays.* Wright-Patterson Air Force Base, OH: Wright Air Development Center.

Cooper, W. E., & Ross, J. R. (1975). World Order. In R. E. Grossman, L. J. San, & T. J. Vance (Eds.), *Papers from the parasession on functionalism* (pp. 63–111). Chicago, IL: Chicago Linguistic Society.

Coury, B. G., Boulette, M. D., & Smith, R. A. (1989). Effect of uncertainty and diagnosticity on classification of multidimensional data with integral and separable displays of system status. *Human Factors, 31*, 551–569.

Croxton, F. E., & Stein, H. (1932). Graphic comparisons by bars, squares, circles, and cubes. *Journal of the American Statistical Association, 27*, 54–60.

Croxton, F. E., & Stryker, R. E. (1927). Bar charts versus circle diagrams. *Journal of the American Statistical Association, 22*, 473–482.

Cuff, D. J. (1973). Colour on temperature maps. *Cartographic Journal, 10*, 17–21.

Culbertson, H. M., & Powers, R. D. (1959). A study of graph comprehension difficulties. *Audio Visual Communication Review, 7*, 97–100.

Curcio, F. R. (1987). Comprehension of mathematical relationships expressed in graphs. *Journal for Research in Mathematics Education, 18*, 382–393.

De Rossi, L. C. (2002). *MasterView International; Creating and Managing Effective PowerPoint Presentations for International Audiences.* Owner: S. Luchini, IKONOS New Media. Created: 15 January 2002; Retrieved 11 December 2005. http://www.masterviews.com/archive/masterview8.htm#serif

DeSanctis, G. (1984). Computer graphics as decision aids: Direction for research. *Decision Science, 15*, 463–487.

DeSanctis, G., & Jarvenpaa, S. L. (1985). An investigation of the "tables versus graphs" controversy in a learning environment. In L. Gallegos, R. Welke, & J. Wetherbe (Eds.), *Proceedings of the 6th international conference on information systems* (pp. 134–144), Indianapolis, IN.

De Soete, G., & De Corte, W. (1985). On the perceptual salience of features of Chernoff faces for representing multivariate data. *Applied Psychological Measurement, 9*, 275–280.

De Valois, R. L., & De Valois, K. K. (1988). *Spatial vision.* New York: Oxford University Press.

Dodwell, P. C. (1975). Perceptual structure and selection. In E. C. Carterette & M. P. Friedman (Eds.), *Handbook of perception* (Vol. 5, pp. 267–300). New York: Academic Press.

Dreyfuss, H. (1984). *Symbol sourcebook: An authoritative guide to international graphic symbols.* New York: John Wiley and Sons.

Dunn, R. (1987). Variable-width framed rectangle charts for statistical mapping. *American Statistician, 41*, 153–156.

Dunn, R. (1988). Framed rectangle charts or statistical maps with shading. *American Statistician, 42*, 123–129.

Eells, W. C. (1926). The relative merits of circles and bars for representing component parts. *Journal of the American Statistical Association, 21*, 119–132.

Ehrenberg, A. S. C. (1975). *Data reduction: Analyzing and interpreting statistical data.* New York: Wiley.

Ericsson, K. A., Chase, W. G., & Faloon, S. (1980). Acquisition of a memory skill. *Science, 208*, 1181–1182.

Feeney, A. & Webber, L. J. (2003). Analogical representation and graph comprehension. In A. Butz, A. Krüger, & P. Olivier (Eds.) *Smart Graphics 2003, Lecture Notes in Computer Science* (Vol. 2733, pp. 212–221). New York: Springer.

Feinberg, S. E. (1979). Graphical methods in statistics. *American Statistician, 33,* 165–178.

Fischer, M. H. (2000). Do irrelevant depth cues affect the comprehension of bar graphs? *Applied Cognitive Psychology, 14,* 151–162.

Fisher, H. T. (1982). *Mapping information.* Cambridge, MA: Abt Books.

Fisher, S. H., Dempsey, J. V., & Marousky, R. T. (1997). Data visualization: Preferences and use of two-dimensional and three-dimensional graphs. *Social Science Review, 15,* 256–263.

Fleming, M. L., & Levie, W. H. (1978). *Instructional message design: Principles from the behavioral sciences.* Englewood Cliffs, NJ: Educational Technology Publications.

Frenckner, K. (1990). *Legibility of continuous text on computer screens—a guide to the literature* (TRITA-NA P9010, IPLab-25). Stockholm: Royal Institute of Technology

Friel, S. N., Curcio, F. R., & Bright, G. W. (2001). Making sense of graphs: Critical factors influencing comprehension and instructional implications. *Journal for Research in Mathematics Education, 32,* 124–158.

Frisby, J. P. (1980). *Seeing: Illusion, brain, and mind.* New York: Oxford University Press.

Garner, W. R. (1970). The stimulus in information processing. *American Psychologist, 25,* 350–358.

Garner, W. R. (1974). *The processing of information and structure.* Hillsdale, NJ: Erlbaum.

Garner, W. R. (1976). Interaction of stimulus dimensions in concept and choice processes. *Cognitive Psychology, 8,* 98–123.

Gattis, M., & Holyoak, K. J. (1996). Mapping conceptual to spatial relations in visual reasoning. *Journal of Experimental Psychology: Learning, Memory, & Cognition, 22,* 231–239.

Geyer, L. H., & DeWald, C. G. (1973). Feature lists and confusion matrices. *Perception & Psychophysics, 14,* 471–482.

Gillan, D. J. (1995). Visual arithmetic, computational graphics, and the spatial metaphor. *Human Factors, 37,* 766–780.

Gillan, D. J., & Callahan, A. B. (2000). A componential model of human interaction with graphs: VI. Cognitive engineering of pie graphs. *Human Factors, 42,* 566–591.

Gillan, D. J., & Lewis, R. (1994). A componential model of human interaction with graphs: 1. Linear regression modeling. *Human Factors, 36,* 419–440.

Gillan, D. J., & Richman, E. H. (1994). Miminalism and the syntax of graphs. *Human Factors, 36,* 619–644.

Glass, A. R., & Holyoak, K. J. (1986). *Cognition* (2nd ed.). New York: Random House.

Goldsmith, T. E., & Schvaneveldt, R. W. (1985). Facilitating multiple-cue judgments with integral information displays. In J. Thomas & M. Schneider (Eds.), *Human factors in computer systems* (pp. 243–270). Norwood, NJ: Ablex.

Gould, J. D., Alfaro, L., Finn, R., Haupt, B., & Minuto, A. (1987). Reading from CRT displays can be as fast as reading from paper. *Human Factors, 29,* 497–517.

Graham, J. L. (1937). Illusory trends in the observation of bar graphs. *Journal of Experimental Psychology, 20,* 597–608.

Gregory, R. L. (1966). *Eye and brain: The psychology of seeing.* New York: McGraw-Hill.

Gregory, R. L. (1970). *The intelligent eye.* London: Weidenfeld & Nicholson.

Grice, H. P. (1975). Logic and conversation. In P. Cole & J. L. Morgan (Eds.), *Syntax and semantics: Vol. 3. Speech acts* (pp. 41–58). New York: Seminar Press.

Hallock, J. (2003). *Color assignment.* Owner: J. Hallock. Created 3 March 2003. Retrieved 12 December 2005. http://www.joehallock.com/edu/COM498/index.html

Hastie, R., Hammerlie, O., Kerwin, J., Croner, C. M., & Herrmann, D. J. (1996). Human performance reading statistical maps. *Journal of Experimental Psychology: Applied, 2,* 3–16.

Helander, M. G., Billingsley, P. A., & Schurick, J. M. (1984). An evaluation of human factors research on visual display terminals in the workplace. In F. A. Muckler (Ed.), *The human factors review* (pp. 55–129). Santa Monica, CA: Human Factors Society.

Held, R. (1980). The rediscovery of adaptability in the visual system: Effects of extrinsic and intrinsic chromatic dispersion. In C. S. Harris (Ed.), *Visual coding and adaptability* (pp. 69–94). Hillsdale, NJ: Erlbaum.

Hicks, M., O'Malley, C., Nichols, S., & Anderson, B. (2003). Comparison of 2D and 3D representations for visualising telecommunication usage. *Behaviour & Information Technology, 22,* 185–201.

Hill, A. (1997). *Readability of websites with various foreground/background color combinations, font types and word styles.* Owner: A. Hill. Created: 1997. Retrieved 15 December 2005. http://hubel.sfasu.edu/research/AHNCUR.html#Fig1

Hitt, W. D. (1961). An evaluation of five different abstract coding methods—experiment IV. *Human Factors, 3,* 120–130.

Hochberg, J. E. (1964). *Perception.* Englewood Cliffs, NJ: Prentice-Hall.

Hollands, J. G. (2003). The classification of graphical elements. *Canadian Journal of Experimental Psychology, 57,* 38–47.

Hollands, J. G., & Spence, I. (1998). Judging proportion with graphs: The summation model. *Applied Cognitive Psychology, 12,* 173–190.

Hollands, J. G., & Spence, I. (2001). The discrimination of graphical elements. *Applied Cognitive Psychology, 15,* 413–431.

Holmes, N. (1984). *Designer's guide to creating charts and diagrams.* New York: Watson-Guptill.

Horn, R. E. (1998). *Visual language.* Bainbridge Island, WA: MacroVu, Inc.

Hotopf, W. H. N. (1966). The size-constancy theory of visual illusions. *British Journal of Psychology, 57,* 307–318.

Hotopf, W. H. N., & Hibberd, M. C. (1989). The role of angles in inducing misalignment in the Poggendorf figure. *Quarterly Journal of Experimental Psychology, 41A,* 355–383.

Howard, I. P. (1982). *Human visual orientation.* New York: Wiley.

Huber, P. J. (1987). Experiences with three-dimensional scatterplots. *Journal of the American Statistical Association, 82,* 448–453.

Huff, D. (1954). *How to lie with statistics.* New York: W.W. Norton & Company.

Jacob, R. J. K., Egeth, H. E., & Bevan, W. (1976). The face as a data display. *Human Factors, 18,* 189–200.

Jarvenpaa, S. L., & Dickson, G. W. (1988). Graphics and managerial decision making: Research based guidelines. *Communications of the ACM, 31,* 764–774.

Jenks, C. F., & Knos, D. S. (1961). The use of shading patterns in graded series. *Annals of the Association of American Geographers, 51,* 316–334.

Johnson, J. R., Rice, R. R., & Roemmich, R. A. (1980). Pictures that lie: The abuse of graphs in annual reports. *Management Accounting, 62,* 50–56.

Jolicoeur, P. (1990). Identification of disoriented objects: A dual-systems theory. *Mind & Language, 5,* 387–410.

Jones, P., & Wickens, C. D. (1986). *The display of multivariate information: The effects of auto- and cross-correlation, display format, and reliability.* Champaign, IL: Cognitive Psychophysiology Laboratory, University of Illinois.

Julez, B. (1980). Spatial-frequency channels in one-, two-, and three-dimensional vision: Variations on a theme by Bekesy. In C. S. Harris (Ed.), *Visual coding and adaptability* (pp. 263–316). Hillsdale, NJ: Erlbaum.

Kahneman, D. (1973). *Attention and effort.* Englewood Cliffs, NJ: Prentice-Hall.

Kaufman, L. (1974). *Sight and mind: An introduction to visual perception.* New York: Oxford.

King, W. C., Dent, M. M., & Miles, E. W. (1991). The persuasive effect of graphics in computer-mediated communication. *Computers in Human Behavior, 7,* 269–279.

Kolata, G. (1984). The proper display of data. *Science, 226*, 156–157.

Kosslyn, S. M. (1976). Using imagery to retrieve semantic information: A developmental study. *Child Development, 47*, 434–444.

Kosslyn, S. M. (1983). *Ghosts in the mind's machine.* New York: W. W. Norton.

Kosslyn, S. M. (1985). Graphics and human information processing: A review of five books. *Journal of the American Statistical Association, 80*, 499–512.

Kosslyn, S. M. (1989). Understanding charts and graphs. *Applied Cognitive Psychology, 3*, 185–225.

Kosslyn, S. M. (1994). *Image and brain: The resolution of the imagery debate.* Cambridge, MA: MIT Press.

Kosslyn, S. M. & James, P. (1980). *Fitting the graph to the question.* Harvard University manuscript unpublished.

Kosslyn, S. M., & Koenig, O. (1995). *Wet mind: The new cognitive neuroscience.* New York: Free Press.

Lauer, T. W., & Post, G. V. (1989). Density in scatterplots and the estimation of correlation. *Behavior Information Technology, 8*, 235–244.

Lederman, S. J., & Campbell, J. I. (1982). Tangible graphs for the blind. *Human Factors, 24*, 85–100.

Lederman, S. J., & Campbell, J. I. (1983). Tangible line graphs: An evaluation and some systematic strategies for exploration. *Journal of Visual Impairment & Blindness, 77*, 108–112.

Legge, G. E., Gu, Y., & Luebker, A. (1989). Efficiency of graphical perception. *Perception & Psychophysics, 46*, 365–374.

Levy, E., Zacks, J., Tversky, B., & Schiano, D. (1996). Gratuitous graphics? Putting preferences in perspective. In M. J. Tauber (Ed.), *Proceedings of the ACM Conference on Human Factors in Computing Systems* (pp. 42–49). Vancouver: ACM.

Lewandowsky, S., & Spence, I. (1989). Discriminating strata in scatterplots. *Journal of the American Statistical Association, 84*, 682–688.

Lohse, G. L. (1997). The role of working memory on graphical information processing. *Behaviour & Information Technology, 16*, 297–308.

Lowe, R. K. (1993). Constructing a mental representation from an abstract technical diagram. *Learning & Instruction, 3*, 157–179.

Lucas, H. C. J. (1981). An experimental investigation of the use of computer-based graphics in decision making. *Management Science, 27*, 757–768.

Macdonald-Ross, M. (1977). How numbers are shown: A review of research on the presentation of quantitative data in texts. *Audio-Visual Communication Review, 25*, 359–409.

MacGregor, D., & Slovic, P. (1986). Graphic representation of judgmental information. *Human-Computer Interaction, 2*, 179–200.

Maclean, I. E., & Stacey, B. G. (1971). Judgment of angle size: An experimental appraisal. *Perception & Psychophysics, 9*, 499–504.

MacLeod, V. M. (1991). Half a century of research on the Stroop effect: An integrative review. *Psychological Bulletin, 109*, 163–203.

Mansfield, J. S., Legge, G. E., & Bane, M. C. (1996). Psychophysics of reading XV: Font effects in normal and low vision. *Investigative Ophthalmology & Visual Science, 37*, 1492–2015.

Marshall, S. (2001). Creating good web sites. Owner: S. Marshall. Created: 2001. Retrieved 12 December 2005, http://www.leafdigital.com/class/lessons/graphicdesign1/4.html

Meihoefer, H. J. (1969). The utility of the circle as an effective cartographic symbol. *Canadian Cartographer, 6*, 105–117.

Meihoefer, H. J. (1973). The visual perception of the circle in thematic maps: Experimental results. *Canadian Cartographer, 10*, 63–84.

Meyer, J. (2000). Performance with tables and graphs: effects of training and a visual search model. *Ergonomics, 43,* 1840–1865.

Meyer, J., Shamo, K., & Gopher, D. (1999). Information structure and the relative efficacy of tables and graphs. *Human Factors, 41,* 570–587.

Meyer, J., & Shinar, D. (1992). Estimating correlations from scatter plots. *Human Factors, 34,* 335–349.

Meyer, J., Shinar, D., & Leiser, D. (1997). Multiple factors that determine performance with tables and graphs. *Human Factors, 39,* 268–286.

Miller, G. A. (1956). The magical number seven, plus or minus two: Some limits on our capacity for processing information. *Psychological Review, 63,* 81–97.

Miller, R. G. (1985). Discussion of "Projection pursuit," by P. J. Huber. *Annals of Statistics, 13,* 510–513.

Milroy, R., & Poulton, E. C. (1978). Labelling graphs for improved reading speed. *Ergonomics, 21,* 55–61.

Mishkin, M., & Appenzeller, T. (1987). The anatomy of memory. *Scientific American, 256,* 80–89.

Mokros, J. R., & Tinker, R. F. (1987). The impact of microcomputer-based labs on children's ability to interpret graphs. *Journal of Research in Science Teaching, 24,* 369–383.

Monmonier, M. (1991). *How to lie with maps.* Chicago, IL: University of Chicago Press.

Moriarity, S. (1979). Communicating financial information through multidimensional graphics. *Journal of Accounting Research, 17,* 205–224.

Morrison, S., & Noyes, J. (2003). *A comparison of two computer fonts: Serif versus Ornate Sans Serif.* Owner: Usability News 5.2. Created 2003. Retrieved 12 December 2005. http://psychology.wichita.edu/surl/usabilitynews/52/UK_font.htm

Mosteller, F., Siegel, A. F., Trapido, E., & Youtz, C. (1985). Fitting straight lines by eye. In D. C. Hoaglin, F. Mosteller, & J. W. Tukey (Eds.), *Exploring data tables, trends, and shapes* (pp. 225–240). New York: John Wiley.

Moxley, R. (1983). Educational diagrams. *Instructional Science, 12,* 147–160.

Nawrocki, L. H. (1973). *Graphic versus tote display of information in a simulated tactical operations system* (Tech. Res. Note No. 243). Washington, DC: Army Research Institute for the Behavioral and Social Sciences.

Neisser, U. (1967). *Cognitive psychology.* New York: Appleton-Century-Crofts.

Neisser, U. (1976). *Cognition and reality.* San Francisco: W. H. Freeman.

Neurath, A. (1974). Isotype. *Instructional Science, 3,* 127–150.

Nothdurft, H. C. (1991). Texture segmentation and pop-out from orientation contrast. *Vision Research, 31,* 1073–1078.

Oron-Gilad, T., Meyer, J., & Gopher, D. (2001). Monitoring dynamic processes with alphanumeric and graphic displays. *Theoretical Issues in Ergonomic Science, 2,* 368–389.

Osherson, D., Kosslyn, S. M., & Hollerbach, J. (Eds.). (1990). *An invitation to cognitive science: Visual cognition and action* (Vol. 2). Cambridge, MA: MIT Press.

Osherson, D. N., & Smith, E. E. (Eds.). (1990). *Thinking: An invitation to cognitive science* (Vol. 3). Cambridge, MA: MIT Press.

Patterson, D. G., & Tinker, M. A. (1940). *How to make type readable: A manual for typographers, printers and advertisers.* New York: Harper & Brothers.

Peebles, D., & Cheng, P. C.-H. (2003). Modeling the effect of task and graphical representation on response latency in a graph reading task. *Human Factors, 45,* 28–46.

Peterson, L. V., & Schramm, W. (1954). How accurately are different kinds of graphs read? *Audio-Visual Communication Review, 2,* 178–189.

Peterson, R. J., Banks, W. W., & Gertman, D. I. (1982). Performance-based evaluation of graphic displays for nuclear power plant control rooms. In J. A. Nichols & M. L. Schneider (Eds.), *Proceedings of the SIGCHI conference on human factors in computing systems* (pp. 182–189). March 15–17, Gaithersburg, Maryland, United States.

Phillips, R. J., Coe, B., Kono, E., Knapp, J., Barrett, S., Wiseman, G., et al. (1990). An experimental approach to the design of cartographic symbols. *Applied Cognitive Psychology, 4,* 485–497.

Pinker, S. (1990). A theory of graph comprehension. In R. Freedle (Ed.), *Artificial intelligence and the future of testing* (pp. 73–126). Hillsdale, NJ: Lawrence Erlbaum Associates.

Pinker, S., & Birdsong, D. (1979). Speakers' sensitivity to rules of frozen word order. *Journal of Verbal Learning & Verbal Behavior, 18,* 497–508.

Playfair, W. (1786). *The commercial and political atlas.* London: Corry.

Pokorny, J., Smith, V. C., Verriest, G., & Pinckers, A. J. L. G. (1979). *Congenital and acquired colour vision defects.* New York: Grune & Stratton.

Pomerantz, J. R. (1981). Perceptual organization in information processing. In M. Kubovy & J. R. Pomerantz (Eds.), *Perceptual organization* (pp. 141–180). Hillsdale, NJ: Erlbaum.

Posner, M. I. (1978). *Chronometric explorations of mind.* Hillsdale, NJ: Lawrence Erlbaum.

Poulton, E. C. (1985). Geometric illusions in reading graphs. *Perception &Psychophysics, 37,* 543–548.

Powell, J. E. (1990). *Designing user interfaces.* San Marcos, CA: Microtrend Books.

Reaven, G. M., & Miller, R. G. (1979). An attempt to define the nature of chemical diabetes using a multidimensional analysis. *Diabetologia, 16,* 17–24.

Reed, S. K., & Johnsen, J. A. (1975). Detection of parts in patterns and images. *Memory & Cognition, 3,* 569–575.

Reid, R. C. (1999). Vision. In M. J. Zigmond, F. E. Bloom, S. C. Landis, J. L. Roberts, & L. R. Squire (Eds.), *Fundamental neuroscience* (pp. 821–851). New York: Academic Press.

Risden, K., Czerwinski, M., Munzner, T., & Cook, D. (2000). An initial examination of ease of use for 2D and 3D information visualizations of web content. *International Journal of Human-Computer Studies, 53,* 695–714.

Robinson, J. O. (1972). *The psychology of visual illusion.* London: Hutchinson.

Sanderson, P. M., Flach, J. M., Buttigieg, M. A., & Casey, E. J. (1989). Object displays do not always support better integrated task performance. *Human Factors, 31,* 183–198.

Sanfey, A., & Hastie, R. (1998). Does evidence presentation format affect judgment? An experimental evaluation of displays of data for judgments. *Psychological Science, 9,* 99–103.

Schacter, D. L. (1996). *Searching for Memory: the brain, the mind, and the past.* New York: Basic Book.

Schiano, D. J., & Tversky, B. (1992). Structure and strategy in encoding simplified graphs. *Memory & Cognition, 20,* 12–20.

Schmid, C. F. (1954). *Handbook of graphic presentation.* New York: Ronald Press Co.

Schutz, H. G. (1961a). An evaluation of formats for graphic trend displays—experiment II. *Human Factors, 3,* 99–107.

Schutz, H. G. (1961b). An evaluation of methods for presentation of graphic multiple trends—experiment III. *Human Factors, 31,* 108–119.

Schwartz, D. R. (1984). *Optional stopping performance under graphic and numeric CRT formatting* (Tech. Rep. No. 84–1). Houston, TX: Department of Psychology, Rice University.

Shah, P. (2002). Graph comprehension: The role of format, content and individual differences. In M. Anderson, B. Meyer, & P. Olivier (Eds.), *Diagrammatic representation and reasoning* (pp. 173–185). Berlin: Springer.

Shah, P., & Carpenter, P. A. (1995). Conceptual limitations in comprehending line graphs. *Journal of Experimental Psychology: General, 124,* 43–61.

Shah, P., Hegarty, M., & Mayer, R. E. (1999). Graphs as aids to knowledge construction: Signaling techniques for guiding the process of graph comprehension. *Journal of Educational Psychology, 91,* 690–702.

Shah, P., & Hoeffner, J. (2002). Review of graph comprehension research: Implications for instruction. *Educational Psychology Review, 14,* 47–69.

Shannon, C. E. (1948). A mathematical theory of communication. *Bell System Technical Journal, 27,* 379–423, 623–656.

Siegrist, M. (1996). The use or misuse of three-dimensional graphs to represent lower-dimensional data. *Behaviour & Information Technology, 15,* 96–100.

Simcox, W. A. (1983). *A perceptual analysis of graphic information processing.* Unpublished doctoral dissertation, Tufts University.

Simcox, W. A. (1984). A method for pragmatic communication in graphic displays. *Human Factors, 26,* 483–487.

Simkin, D., & Hastie, R. (1987). An information processing analysis of graph perception. *Journal of the American Statistical Association, 82,* 454–465.

Smallman, H. S., & Boyton, R. M. (1990). Segregation of basic colors in an information display. *Journal of the Optical Society of America, A7,* 1985–1994.

Smith, E. E., & Kosslyn, S. M. (2006). *Cognitive psychology: Mind and brain.* New York: Prentice Hall.

Smith, S. L. (1979). Letter size and legibility. *Human Factors, 21*(b), 661–670.

Sparrow, J. A. (1989). Graphical displays in information systems: Some data properties influencing the effectiveness of alternate forms. *Behaviour & Information Technology, 8,* 43–56.

Spence, I. (1990). Visual psychophysics of simple graphical elements. *Journal of Experimental Psychology: Human Perception & Performance, 16,* 683–692.

Spence, I., & Lewandowsky, S. (1991). Displaying proportions and percentages. *Applied Cognitive Psychology, 5,* 61–77.

Stevens, S. S. (1974). Perceptual magnitude and its measurement. In E. C. Carterette & M. P. Friedman (Eds.), *Handbook of perception* (Vol. 2, pp. 361–389). New York: Academic Press.

Stevens, S. S. (1975). *Psychophysics.* New York: Wiley.

Stock, D., & Watson, C. J. (1984). Human judgment accuracy, multidimensional graphics, and humans versus models. *Journal of Accounting Research, 22,* 192–206.

Teghtsoonian, M. (1965). The judgment of size. *American Journal of Psychology, 78,* 392–402.

Tinker, M. A. (1963). *Legibility of print.* Ames, IA: Iowa State University Press.

Tractinsky, N., & Meyer, J. (1999). Chartjunk or goldgraph? Effects of presentation objectives and content desirability on information presentation. *MIS Quarterly, 23,* 397–420.

Travis, D. (1991). *Effective color displays: Theory and practice.* New York: Academic Press.

Trickett, S. B., Ratwani, R. J., and Trafton, J. G. (2005). *Real-world graph comprehension: High-level questions, complex graphs, and spatial cognition.* Unpublished manuscript, George Mason University.

Trumbo, B. E. (1981). A theory for coloring bivariate statistical maps. *American Statistician, 35,* 220–226.

Tufte, E. R. (1983). *The visual display of quantitative information.* Cheshire, CT: Graphics Press.

Tufte, E. R. (1990). *Envisioning information.* Cheshire, CT: Graphics Press.

Tukey, J. W. (1972). Some graphic and semigraphic displays. In T. A. Bancroft (Ed.), *Statistical papers in honor of George W. Snedecor* (pp. 293–316). Ames: Iowa State University Press.

Tukey, J. W. (1977). *Exploratory data analysis.* Reading, MA: Addison-Wesley.

Tullis, T. S. (1981). An evaluation of alphanumeric, graphic, and colour information displays. *Human Factors*, 23, 541–550.

Tullis, T. S., Boynton, J. L., & Hersch, H. (1995). Readability of fonts in the windows environment. In *Proceedings of CHI'95* (pp. 127–128). Denver, CO: ACM Press.

Tversky, B., Morrison, J. B., & Betrancourt, M. (2002). Animation: Does it facilitate? *International Journal of Human-Computer Studies*, 57, 247–262.

Tversky, B., & Schiano, D. J. (1989). Perceptual and cognitive factors in distortions in memory for graphs and maps. *Journal of Experimental Psychology: General*, 118, 387–398.

Tversky, B., Zacks, J., Lee, P. U., & Heiser, J. (2000). Lines, blobs, crosses, and arrows: Diagrammatic communication with schematic figures. In M. Anderson, P. Cheng, & V. Haarslev (Eds.), *Theory and application of diagrams* (pp. 221–230). Berlin: Springer.

Umanath, N. S., & Scamell, R. W. (1988). An experimental evaluation of the impact of data display format on recall performance. *Communications of the ACM*, 31, 562–570.

Umanath, N. S., & Vessey, I. (1994). Multiattribute data presentation and human judgment: A cognitive fit perspective. *Decision Sciences*, 25, 795–824.

Verdi, M. P., & Kulhavy, R. W. (2002). Learning with maps and texts: An overview. *Educational Psychology Review*, 14, 27–46.

Vernon, M. D. (1946). Learning from graphic material. *British Journal of Psychology*, 36, 145–158.

Vessey, I. (1991). Cognitive fit: A theory-based analysis of the graphs vs. tables literature. *Decision Sciences*, 22, 219–241.

Von Huhn, R. (1927). Further studies in the graphic use of circles and bars. *Journal of the American Statistical Association*, 22, 31–36.

Vos, J. J. (1960). Some new aspects of color stereoscopy. *Journal of the Optical Society of America*, 50, 785–790.

Wainer, H. (1979). *The Wabbit: An alternative icon for multivariate data display* (BSSR Tech. Rep. 547–792). Washington, DC: Bureau of Social Science Research.

Wainer, H. (1992). Understanding graphs and tables. *Educational Researcher*, 21, 14–23.

Wainer, H., & Francolini, C. M. (1980). An empirical inquiry concerning human understanding of two-variable color maps. *American Statistician*, 34, 81–93.

Wainer, H., & Thissen, D. (1981). Graphical data analysis. *Annual Review of Psychology*, 32, 191–241.

Wakimoto, K. (1977). A trial of modification of the face graph proposed by Chernoff: Body graph. *Quantitative Behavioral Science, Kodo Keiryogaku*, 4, 67–73.

Washburne, J. N. (1927). An experimental study of various graphic, tabular and textural methods of presenting quantitative material. *Journal of Educational Psychology*, 18, 361–376, 465–476.

Wheildon, C. (1995). *Type and layout: How typography and design can get your message across—or get in the way*. Berkeley, CA: Strathmore Press.

Wickens, C. D., & Andre, A. D. (1990). Proximity compatibility and information display: Effects of color, space, and objectness on information integration. *Human Factors*, 32, 61–77.

Wickens, C. D., & Carswell, C. M. (1995). The proximity compatibility principle: Its psychological foundation and relevance to display design. *Human Factors*, 37, 473–494.

Wickens, C. D., & Hollands, J. G. (2000). *Engineering psychology and human performance* (3rd ed.). Upper Saddle River, NJ: Prentice Hall.

Williams, R. (2003). *The Mac is not a typewriter: A style manual for creating professional-level type on your Macintosh*, 2nd ed. Berkeley, CA: Peachpit Press.

Wilson, R. F. (2001). *HTML E-Mail: Text font readability study*. Owned by *Web Mar-

keting Today, 97. Created 2001. Retrieved 12 December 2005. http://www.wilsonweb.com/wmt6/html-email-fonts.htm

Winn, B. (1987). Charts, graphs, and diagrams in educational materials. In D. M. Willows & H. A. Houghton (Eds.), *The psychology of illustration: Vol. 1. Basic research* (pp. 152–198). New York: Springer-Verlag.

Winn, W. D. (1983). Perceptual strategies used with flow diagrams having normal and unanticipated formats. *Perceptual & Motor Skills, 57*, 751–762.

Yager, D., Aquilante, K., & Plass, R. (1998). High and low luminance letters, acuity reserve, and font effects on reading speed. *Vision Research, 38*, 2527–2531.

Zachrisson, B. (1965). *Studies in the legibility of printed text.* Stockholm, Sweden: Almqvist & Wiksell.

Zacks, J., & Tversky, B. (1999). Bars and lines: A study of graphic communication. *Memory & Cognition, 27*, 1073–1079.

Zacks, J., Levy, E., Tversky, B., & Schiano, D. (2001). Graphs in use. In M. Anderson, B. Meyer, & P. Olivier (Eds.), *Diagrammatic reasoning and representation* (pp. 187–206). Berlin: Springer.

Zacks, J., Levy, E., Tversky, B., & Schiano, D. J. (1998). Reading bar graphs: Effects of extraneous depth cues and graphical context. *Journal of Experimental Psychology: Applied, 4*, 119–138.

Sources of Data and Figures

All of the figures that appear in this book are original drawings, prepared by Network Graphics, Vantage Art, and Megan Higgins. Many use fictional data; others, listed below, were adapted from the following sources:

9 "The recognition of faces" by L. D. Harmon, 1973, *Scientific American*, 229, p. 75.

12 *left*, Consensus Economics, Inc., London, cited in *The Economist*, 15 December 1990, p. 99; *right*, *The Economist*, 4 May 1991, p. 103.

24 "Social policy and recent fertility changes in Sweden" by J. M. Hoem, *Population and Development Review*, December 1990, cited in *The Economist*, 13 April 1991, p. 47.

39 U.S. Federal Reserve and Bureau of Labor Statistics, 2.5 cited in *The Economist*, 1 December 1990, p. 94.

41 U.S. Federal Reserve and Bureau of Labor Statistics, 2.5 cited in *The Economist*, 1 December 1990, p. 94.

43 *top*, DRI/McGraw-Hill, cited in *The Economist*, 8 June 1991, Survey p. 23; *bottom*, ASEA Annual Report, 1989.

44 U.S. Forest Service, cited in *The Economist*, 22 June 1991, p. 23.

55 Inland Revenue (U.K.), cited in *The Economist*, 14 March 1992, p. 73.

57 Kosslyn, S. M., Sokolov, M. A., & Chen, J. C. (1989). The lateralization of BRIAN: A computational theory and model of visual hemispheric specialization. In D. Klahr and K. Kotovsky (Eds.), *Complex information processing: The impact of Herbert H. Simon*. Hillsdale, NJ: Erlbaum.

59 DRI/McGraw-Hill, cited in *The Economist*, 8 June 1991, p. 23.

61 *Consumer Reports*, April 1979, and U.S. Environmental Protection Agency, cited in Chambers, Cleveland, Kleiner, & Tukey (1983, p. 139).

62 *Consumer Reports*, April 1979, and U.S. Environmental Protection Agency, cited in Chambers et al. (1983, p. 133).

66 *The Scientist*, 5 March 1990, p. 10.

70 *Petrology* (p. 295) by E. G. Ehlers and H. Blatt, 1982, New York: W. H. Freemen.

72 Cleveland (1985, p. 276).

81 *The New York Times*, 20 October 1991, sec. 8, p. 2.

86 The Royal Society, cited in *The Economist*, 18 May 1991, p. 63.

87 Leo J. Shapiro & Associates, cited in *The Wall Street Journal*, 4 October 1991, p. B1.

89 *The computational brain: Models and methods on the frontiers of computational neuroscience* (p. 428) by P. S. Churchland and T. J. Sejnowski, 1992, Cambridge, MA: MIT Press.

90 U.S. Immigration and Naturalization Service, 3.16 cited in *The Economist*, 11 May 1991, p. 18.

91 Arthur Andersen, Inc., cited in *The Economist*, 17 August 1991, p. 67.

92 The Conference Board, cited in *The Economist*, 18 May 1991, p. 74.

95 *The Wall Street Journal*, 21 May 1990, p. A4.

96 Datastream, U.S. Department of Commerce, and Yamaichi Research Institute, cited in *The Economist*, 31 August 1991, p. 84.

97 Metropolitan Life Insurance Company height and weight tables, cited in *The psychology of eating and drinking: An introduction* (p. 172) by A. W. Logue, 1991, New York: W. H. Freeman.

101 *top, The Economist*, 31 August 1991, p. 17.

105 *top*, Department of the Environment (U.K.), cited in *The Economist*, 20 April 1991, p. 56; *bottom, The Economist*, 14 March 1992, p. 67.

109 World Resources Institute, cited in *Ecology and the politics of scarcity revisited* (p. 74), W. Ophuls and A. S. Boyan, Jr., 1992, New York: W. H. Freeman.

118 *top, Cultural Trends, 1990*, cited in *The Economist*, 15 December 1990, p. 55.

121 National Association of Manufacturers and Bank of England, cited in *The Economist*, 12 January 1991, p. 60.

128 *Jesse Meyers' Beverage Digest*, cited in *Newsweek*, 19 March 1990, p. 38.

129 *bottom, The Economist*, 23 November 1991, Survey p. 13.

130 *The why and how of home horticulture*, second ed. (p. 480), by D. R. Bienz, 1993, New York: W. H. Freeman.

131 *top*, *Jesse Meyers' Beverage Digest*, cited in *Newsweek*, 19 March 1990, p. 38.

133 Greater Tampa Association of Realtors, St. Petersburg Suncoast Association of Realtors, and Greater Clearwater Board of Realtors, cited in *Tampa Tribune*, 29 May 1991.

134 *Cultural Trends, 1990*, cited in *The Economist*, 15 December 1990, p. 55.

136 *top*, *The evolution of Japanese direct investment in Europe* by S. Thomsen and P. Nicolaides, 1991, New York: Harvester Wheatsheaf, and U.S. Department of Commerce, cited in *The Economist*, 20 April 1991, p. 65.

137 *The beginnings of social understanding* (p. 48) by J. Dunn, 1988, Cambridge, MA: Harvard University Press, cited in *The development of children*, second ed. (p. 377), by M. Cole and S. R. Cole, 1993, New York: W. H. Freeman.

139 *The sports encyclopedia: pro football* by David Neft, Richard Cohen, and Rick Korch, 1992, New York: St. Martin's Press.

142 Nielsen Media Research, cited in *Newsweek*, 26 March 1990, p. 58

143 Schutz (1961b).

147 Dun & Bradstreet and Department of Employment (U.K.), cited in *The Economist*, 5 January 1991, p. 42.

148 The Economist, 23 November 1991, p. 59

149 Greater Tampa Association of Realtors, St. Petersburg Suncoast Association of Realtors, and Greater Clearwater Board of Realtors, cited in *Tampa Tribune*, 29 May 1991.

151 Department of the Environment (U.K.), cited in *The Economist*, 20 April 1991, p. 56.

160 *The new archaeology and the ancient Maya* (p. 130) by J. A. Sabloff, 1990, New York: Scientific American Library.

162 "Development of moral judgement: A longitudinal study of males and females" by C. Holstein, 1976, *Child Development*, 47, pp. 51–61, cited in Cole & Cole (1993, p. 629).

172 King's Fund Institute, cited in *The Economist*, 22 February 1992, p. 49.

180 Baseball Commissioner, cited in *The Economist*, 13 April 1991, p. 67.

181 *bottom*, *Fortune* and *Forbes* magazines, cited in *Designer's guide to creating charts and diagrams* (p. 46) by N. Holmes, 1984, New York: Watson-Guptill.

183 *top*, Comité Interprofessionnal du Vin de Champagne, cited in *The Economist*, 1 December 1990, p. 69.

184 U.S. Department of Commerce and Bureau of the Census, cited in *The Economist*, 6 July 1991, p. 26.

185 *The economy of nature*, third ed. (p. 71), by R. E. Ricklefs, 1993, New York: W. H. Freeman.

187 *top*, *The New York Times*, 17 February 1992, p. A12; *bottom*, SIPRI, cited in *The Economist*, 6 July 1991, p. 33.

190 *Infancy: Its place in human development* by J. Kagan, R. B. Kearsley, and P. Zelazo, 1978, Cambridge, MA: Harvard University Press, cited in Cole & Cole (1993, p. 236).

191 *The New York Times*, 26 May 1993, p. B7.

196 UNESCO *UN Demographic Survey* (1982–86 average figures), cited in *The Economist World Atlas and Almanac*, 1989, p. 114.

197 *The New York Times*, 20 October 1991, sec. 4, p. 3.

198 Telerate Teletrac, cited in *The Wall Street Journal*, 2 November 1990, p. C1.

199 Ricklefs (1993, p. 441).

203 Natural Resources Defense Council.

204 Natural Resources Defense Council.

205 *top and bottom left*, Natural Resources Defense Council.
bottom right, Morgan Stanley Capital International, cited in *The Economist World Atlas and Almanac*, 1989, p. 90.

207 Weisberg (1980), cited in Chambers et al. (1983, p. 371).

208 EIU *Country Report*, cited in *The Economist World Atlas and Almanac*, 1989, p. 267.

209 *UN Energy Statistics Yearbook, Demographic Yearbook* (1985 figures), cited in *The Economist World Atlas and Almanac*, 1989, p. 95.

210 UNESCO *UN Demographic Survey* (1982–86 average figures), cited in *The Economist World Atlas and Almanac*, 1989, p. 114.

211 EIU *Country Profile*, cited in *The Economist World Atlas and Almanac*, 1989, p. 162.

212 Morgan Stanley Capital International, cited in *The Economist World Atlas and Almanac*, 1989, p. 90.

214 Morgan Stanley Capital International, cited in *The Economist World Atlas and Almanac*, 1989, p. 90.

215 *top and bottom*, Food and Agricultural Organizations 8.14 (1979–81 average), cited in *The Economist World Atlas and Almanac*, 1989, p. 106.

217 Weisberg (1980), cited in Chambers et al. (1983, p. 371).

218 Weisberg (1980), cited in Chambers et al. (1983, 8.19 p. 371).

219 *Two nations: Black and white, separate, hostile, unequal* (p. 98) by A. Hacker, 1992, New York: Scribner's, cited in *Newsweek*, 23 March 1992, p. 61.

220 U.S. Department of Transportation, cited in *The Economist*, 10 March 1990, p. 73.

221 *top*, U.S. Department of Commerce, cited in *Newsweek*, 2 March 1992, p. 38.

p. 38.

222 Kosslyn, S. M., Ball, T. M., & Reiser, B. J. (1978). Visual images preserve metric spatial information: *Evidence from studies of image scanning. Journal of Experimental Psychology: Human Perception and Performance, 4,* 47–60.

223 Kosslyn (1976).

225 U.S. Department of Commerce, cited in *Newsweek,* 23 March 1992, p. 48.

Index